Bandera's Boys:

Twelve Historic Wargames Scenarios based on the actions of the Ukrainian Insurgent Army during and after WWII

Adrian Mandzy

Cover image courtesy of Adrian Mandzy
Edited by Vincent W. Rospond

This edition published in 2023

Winged Hussar Publishing is an imprint of

Winged Hussar Publishing, LLC
1525 Hulse Rd, Unit 1
Point Pleasant, NJ 08742

Copyright © Winged Hussar Publishing
ISBN PB 978-1-958872-24-6
Bibliographical References and Index
1. History. 2. Ukraine. 3. Military.

Winged Hussar Publishing, LLC / Adrian Mandzy All rights reserved
For more information
visit us at www.whpsupplyroom.com

Twitter: WingHusPubLLC
Facebook: Winged Hussar Publishing LLC

This book is sold subject to the condition that it shall not, by way of trade or otherwise, be lent, resold, hired out, or otherwise circulated without the publisher's prior consent in any form of binding or cover other than that in which it is published and without a similar condition, including this condition, being imposed on the subsequent purchaser.

The scanning, uploading, and distribution of this book via the Internet or via any other means without the permission of the publisher is illegal and punishable by law. Please purchase only authorized electronic editions, and do not participate in or encourage electronic piracy of copyrighted materials. Your support of the author's and publisher's rights is appreciated. Karma, it's everywhere.

# Table of Contents

Acknowledgments .................................................................................................... 4

Preface - Philosophical Statement on War and Wargaming ................................................... 5

The UPA - An Introduction ........................................................................................ 8

Background History of the Ukrainian Insurgent Army ......................................................... 9

UPA Organizational Structure .................................................................................... 13

Scenario #1 - Assassination of a High-Ranking Nazi, May 1943 - Klevan ........................ 17

Scenario #2 - UPA attack on a German Military train, 23-24 June 1943 - Nemovychi .... 23

Scenario #3 - Liberation of a Penal Labor Work Camp, July 1943 - Sviatoslav ................ 32

Scenario #4 - Raid on a town, August 1943 - Kamin Koshyrsky ..................................... 40

Scenario #5 - German attack on a UPA NCO school, fall 1943 - Black Forest ................... 48

Scenario #6 - Scenario #6, Hold the Line, April 1944 – Hurby ....................................... 58

Scenario #7 - Winter Mountain Ambush, January 1945 - Kosmach ................................. 64

Scenario #8 - Clash between UPA insurgents and Soviet partisans, February 1945 ......... 70

Scenario #9 - Breakthrough into West Germany - 1947 ................................................ 75

Scenario #10 - Escape from the Bunker, March 1948 ................................................... 82

Scenario #11 - Collect the Drop, 1949 ....................................................................... 88

Appendix #1 - Weapons of the Ukrainian Insurgent Army .............................................. 103

Appendix #2 - Uniforms of the Ukrainian Insurgent Army ............................................. 116

Appendix #3 - Figures for the Ukrainian Insurgent Army .............................................. 124

Appendix #4 - Terms and Acronyms .......................................................................... 127

Appendix #5 – List of Figures and Illustrations ........................................................... 130

Further Reading .................................................................................................... 137

# Acknowledgments

A number of individuals made this guidebook possible and I need to publically acknowledge their contributions. Jurko Datsko and Volodymyr Chornovus provided access to historical materials and photographs while Petro R. Sodol shared his research on the UPA with the author. Discussions with other Ukrainian scholars, such as Volodymyr Viatrovych, Ruslan Zabilyj, and Sviatoslav Sheremeta, along with dialogues with former UPA insurgents Myroslav Symchych and Mykola Kulyk gave the author an insight into the operations and the day-to-day struggle of the Ukrainian underground. Special thanks also go to Volodymyr Shturm for allowing the author to make use of his extensive private collection.

UPA reenactors, both in the USA and in Ukraine, also shared their photographs with the author and I thank them for allowing me to use their images in this work. Special thanks to the Greywolves Co. Reenactment Group for sharing with me their research on UPA uniforms and allowing me to reproduce parts of their research in Appendix 2.

Jon Russell of Warlord US provided a number of Warlord figures to be rebuilt and painted to represent UPA insurgents, while Warlord UK allowed the author to use many of the photographs of their Nazi and Soviet models that appear throughout the book. Jon also shared with me pictures of miniatures from his private collection, one of which is reproduced in this scenario book. A special thanks also goes out to Wargames Atlantic, who contributed a box of their 28mm hard plastic Partisans (1) French Resistance to this project. Fellow gamers Tod Kershner and Karl Shanstrom both read the text, shared photographs, and provided many useful comments. Gamer Matthew Doyle also deserves special recognition, as he introduced me to *Bolt Action* and encouraged me to create a 28mm UPA insurgent army. Thomas Pinette also needs to be recognized as he made numerous suggestions on how the historical actions could be better adopted to for use with Warlord's *Bolt Action* rule set. David Tuck kindly offered help in colorizing the maps provided within this work, but in the end the maps that appear were created by Christian Wright, my former undergraduate research fellow who drew the maps for my earlier scenario book, *Bad Roads and Poor Rations; Fifty-Nine Wargame Scenarios for the North American War of 1812*. I also want to thank Mykyta Karpukhin, who runs a facebook page which displays various Ukrainian historical miniatures, Roman Romashko of Trizub Miniatures, who produce miniatures in various scales and materials, and fellow gamer Casey Pittman. All three individuals provided photographs of their craft, some of which appear below. A final bit of thanks go to my editor, Vincent Rospond, who encouraged me to undertake this volume at a time when Russian tanks were rolling into Ukraine.

Adrian Mandzy

## Preface - Philosophical Statement on War and Wargaming

War is a tragedy. Lives are destroyed. Combatants and non-combatants suffer; many lose their lives. The survivors carry scars and often pass them down through the next generations. Individuals who claim that they are striving to build a better world while in pursuit of their own personal empire often bring untold death, destruction and suffering to the innocent.

The Second World War was no different and in the areas between Berlin and Moscow became what Timothy Snyder referred to as *Bloodlands*.[1] The destruction of the region's Jewish population during the Second World War cannot be sidelined, minimized or denied, as is proposed by some individuals. At the same time, we also need to recognize that other populations suffered and were killed for their ethnic origins, ideology, religious beliefs, or sexual orientation. Familiar with the wide-spread horrors of totalitarian systems, in the years that followed 1945 western democracies publicly embraced the idea of "human rights." Though not always followed and violated on more than one occasion, the western liberal tradition recognizes both the importance of human rights and the idea that war is bad.

Not all societies follow these western ideals and some look at war in a positive way. Russia in the first two decades of the 20$^{th}$ century has embraced the cult of the "Great Patriotic War". Created in an era when the Soviet Union no longer saw itself as revolutionary or even moving forward, the past was re-written to glorify the state. While the west used slogans like "never again", in the early years of the 20$^{th}$ century, the Russian government encouraged its citizens to use the phrase "We can repeat". Such ultra-patriotic flag waiving exercises encouraged the Russian population to embrace war and cheer on Moscow's renewed drive for empire.

What does this have to do with wargaming? While some want to ignore war, hiding one's head in the sand is not a way of decreasing state sponsored violence. Such pacifistic thoughts failed the Ethiopians when the Italians invaded and continue to fail to this day. Our wargames do not prepare us for further real time military conquests, but rather provide a different way of looking at the past. The goal is not to glorify the actions of any political organization or its military apparatus, but to reconnect with the events of the time. As gamers, we often get distracted by pretty uniforms and exotic equipment, but at the end of the day, our plastic and metal toy soldiers get put way safely. The same cannot be said for those who lived through such terrible times of war.

---

[1] Timothy Snyder, *Bloodlands: Europe Between Hitler and Stalin.* New York: Basic Books, 2010.

Figure 1. 28mm UPA figures, converted from Warlord's range of Bolt Action models, author's collection.

The book you are holding in your hand provides the historical wargamer with a practical guide to a not well-known aspect of mid twentieth century warfare. Much of the scholarship on the UPA is not in English, and that which is, rarely focuses on military activity. The historical information presented about the UPA and the military engagements they fought presented in this volume is based on current scholarship. In writing this guide, the author made extensive use of historical documents, photographic evidence, secondary works, and battlefield archaeology.

The engagements described below are only a few of the fights that occurred between the UPA and their adversaries. Eleven of the twelve scenarios are tabletop recreations of well documented actions fought between 1943 and 1954, while the twelfth is a what-if scenario based on historical events. Some of the scenarios included in this guide are representative of the typical operations conducted by the UPA, while others appear because they provide the gamer with interesting tactical opportunities.

The scenarios themselves are not tied to a particular set of rules. Some of the actions are small scale and can be easily adopted to Warlord's *Bolt Action* rule set. Other actions are larger and can be gamed using such rules as *Battlegroup* or *Flames of War*. As the gamer, you should use whichever set of rules you or your group enjoy the most. If something does not feel right, change it. If you are unable to field the troops listed in the scenario game notes, or you do not have the table space, feel free to scale it down. Conversely, one can increase the number of troops based on the historical description of the battle. The game notes are simply suggestions on how to translate historical events onto the tabletop.

# The UPA – an Introduction

Of all territories under Soviet and Nazi occupations in the twentieth century, the Ukrainian nationalists mounted one of the largest and longest-lasting asymmetric movements in their efforts to seek independence. From the 1920s until the early 1960s, individuals dedicated to the cause of an independent Ukrainian state waged military operations against Polish, German, and Soviet authorities. Scholars estimate that the Ukrainian Insurgent Army or *Ukrains'ka Povstancha Armia* (UPA), often referred to as either the Ukrainian Insurgent Army or the Ukrainian Partisan Army, fielded approximately 40,000 combat troops. Perhaps more importantly, the UPA was not supplied by any foreign power and all arms and supplies were either created locally or captured from the enemy. The insurgents received a high level of support from the local populous, and it is estimated that as many as 2,500,000 individuals may have provided support to the UPA. The longevity of the conflict, along with the significant grass-root support these insurgents received from the local population, has led many to idealize the actions of these individuals. As in other national liberation struggles, the various ethnic civilian populations caught in the middle suffered and tragedies took place at a time when mass executions, deportations and mass resettlements were an everyday occurrence. Massacres conducted by any side cannot be justified in any manner, but they do need to be understood within their historical content.

In the intervening seventy years since the UPA stopped operating as an organization, Soviet and now Russian authorities continue to brand individuals who do not wish to submit to Moscow's authority as "Banderites". The term "Banderites" at its core signifies an individual to be a follower of Stepan Bandera (1909-1959), the nationalistic minded Ukrainian leader focused on creating an independent Ukrainian state. During the 2022 Russo-Ukrainian War, the term "Banderite", like the previously used terms of Peltura and Mazepa, has moved beyond its historical confines and has come to embody anyone who desires independence from Moscow.

Figure 2. Stephan Bandera.

Adrian Mandzy
# Background History of the Ukrainian Insurgent Army

The UPA did not arise overnight but evolved from earlier efforts at national liberation. Following the failed attempt to create an independent Ukrainian state in the aftermath of the collapsed Imperial Russian and Austro-Hungarian Empires, Ukrainians found themselves living under various political regimes. The majority of ethnic Ukrainians were incorporated into the Soviet Union, whereas others were ruled by Poland, Czechoslovakia, Hungary and Romania. Uncompromising in their devotion to the ideal of a Ukrainian national state, a small handful of Ukrainian veterans continued to push for political independence.[2] Recognizing that creating and supplying a traditional military force of battalions, regiments, divisions and corps, was extremely difficult when facing a much larger, established, conventional army, these revolutionaries sought other methods to achieve their goal of an independent Ukraine.

Following World War I, the Polish government, eager to reduce the Ukrainian presence on its eastern frontiers, used a variety of tactics, including: restrictions on language and education, forced religious conversion, and distribution of state lands to Polish settlers, to create a monolithic Polish state.[3] Throughout the 1920s, the Ukrainian Military Organization (UVO) employed terrorist tactics modeled after those successfully pioneered by Michael Collins in attaining Ireland's independence from Britain. The Polish regime responded by launching a pacification campaign against various established legal Ukrainian institutions.[4] The conduct of the pacification campaign resulted in more Ukrainians supporting and joining the radical nationalists. Taking advantage of its notoriety among the new post-war generation, the UVO sought to merge the young Ukrainian revolutionaries into a new entity, the Organization of Ukrainian Nationalists (OUN). By the early 1930s the OUN had replaced UVO and its members committed various terrorist acts—assassinations, arson and robberies - against both the Polish and Soviet governments from which they sought independence.[5]

As early as 1934, tensions in the OUN began to surface between older members living abroad and those within Ukraine. With the assassination of Evhen Konovalets, the leader of the OUN by a Soviet agent in 1938, the internal friction grew and the younger generation of resident OUN members began to take on greater responsibilities. In early 1939, OUN volunteers crossed into Czechoslovakia and

---

[2] Wolodymyr Kosyk, *The Third Reich and Ukraine. Studies in Modern European History*, Vol. 8. New York: Peter Lang Publishing, Inc. 1993, p.14.

[3] Borovych, Ukraine and Poland, *Ideia i Chyn* 2, no. 4 (1943). In *Political Thought of the Ukrainian Underground, 1943-1951*. Edited by Peter J. Potichnyj and Yevhen Shtendera. Toronto: Canadian Institute of Ukrainian Studies Press. 1986, 62.

[4] Samuel A. Wallace and Yaroslaw Chyz, Western Ukraine under Polish Yoke: Polonization, Colonization, and "Pacification". New York: *The Ukrainian Review* 1931.

[5] Mykola Posivnych, *Voenno-politychna Diial'nist' OUN v 1929-1939/Military-political Activities of the OUN in 1929-1939*. L'viv: National Academy of Sciences of Ukraine, 2010.

fought for the short lived Transcarpathia. Imprisoned by the invading Hungarians, several OUN members were turned over to Polish authorities who either executed them on the spot or sent them to Polish prisons.[6]

In August 1939, Nazi Germany and the Soviet Union signed the Molotov-Ribbentrop Pact. This non-aggression pact, which was signed in Moscow on 23 August contained a Secret Protocol, which partitioned Poland between the two powers and sanctioned the Soviet Union's annexation of Estonia, Lithuania, Latvia, and parts of Finland.

In early September, Poland was invaded, first by Germany and later that month by the Soviet Union. The Soviet advance into western Ukraine, although touted as "Ukrainian liberation from Polish rule," actually brought little relief for the local inhabitants. Initially, the Soviets dismantled the existing Polish political, socioeconomic, and cultural infrastructure, which had the greatest impact on the middle of the road liberal organizations. By the spring of 1940, mass arrests and deportations of Polish, Jewish, and Ukrainians by Soviet authorities resulted in thousands of alleged "enemies of the people" being packed into cattle cars and sent to work in slave labor camps in Siberia and Kazakhstan.[7]

Figure 3 Evhen Konovalets.

In the wake of the German and Soviet invasion of Poland, many OUN members escaped from prison and renewed their struggle for national liberation. By 1940, the OUN had split into two factions – an older generation dominated by veterans of 1917-20 (OUN-M), and the younger radicals who joined the organization in the 1930s (OUN-B). To find needed support for their resistance to Soviet persecution, both factions of the OUN sought to increase cooperation with the Nazi leadership. Two small OUN military units, *Nachtigall*" and *Roland*, were formed for the upcoming "Crusade against Communism".[8]

---

[6]Iaroslav Onyshchuk, *Doslidzhennia pokhovannia bijtsiv Karpats'koi Ukrainy na Verets'komy perevali v 2011 rotsi/Research of the Burial Places of Carpathian Ukraine's Soldiers on Veretsky Mountain Pass in 2011*. L'viv: *Citadel*. No 9. 2013. pp. 46-50.

[7]Orest Subtelny, *Ukraine, A History*. Toronto: University of Toronto Press, 1991. p. 456.

[8]Andrij Bol'novs'kyj, *Ukrains'ki vijs'kovi formuvannia v zbrojnykh sylakh Nimechchyny (1939-1945)/ Ukrainian Military Formations in the German Armed Forces (1939-1945)*. L'viv: Ivan Franko L'viv National University Press, 2003.

Figure 4. Nazi-Soviet Freindship, Nazi Soviet troops stage a joint Victory Parade in the city of Brest, 1939.

When Germany invaded the USSR in 1941, the retreating Soviets executed as many "undesirables" as they could. The Nazi publicized these executions and began to conduct their own mass executions of civilians. In the ensuing chaos, on June 30, 1941, the younger generation of the OUN proclaimed an independent Ukraine.[9] Other OUN members organized themselves into "expeditionary groups" and moving east with the advancing Germans, began to organize local Ukrainians to create a local administration around them. The Nazis pressured Bandera to rescind the Proclamation of Ukrainian Statehood and when he refused, he was arrested. Eventually Bandera was sent to the Sachsenhausen Concentration Camp and two of his brothers perished at Auschwitz. The German political authorities were angered and responded by detaining and shooting those OUN members they could locate.[10] Those "fortunate" enough to be arrested were sent to concentration camps. The mass killing grounds the Nazis used to liquidate Jews living in Ukraine, such as Babin Yar, was also used by the Germans to exterminate UON members.[11]

Figure 5. *Nachtigall* soldiers in the city of L'viv, 30 June 1941

---

[9]Western Ukraine Declares Its Independence. *Trident*, Vol. V, No. 6, July-August 1941, p.3.

[10]Ilnytzkyj, Roman, ed. *Einsatzkommando Order against the Bandera Movement, dated 25 November 1941. In Deutschland und die Ukraine, 1934-45,* Vol. 2. Munich: Osteuropa-Institut, 1956, pp 338-9.

[11]At the Babin Yar Ravine outside of Kyiv, the Nazis executed more than 33,000 Jews in the summer of 1941. The Nazis used machineguns and pistols to kill the Jews. Those wounded were often buried alive with the dead. During the Nazi occupation, other "subversive" people were also executed by the Germans and their collaborators at Bain Yar, including gypsies, Ukrainians (both nationalists and non-political active civilians), and Soviet prisoners of war. It is estimated by some that more than 100,000 residents of Kyiv were executed by the Nazis at this location. On 1 March 2022, during the Russian invasion of Ukraine, a Russian missel struck Babin Yar and killed five people.

Figure 6. Nazi Executions, 2 December 1943, Drohobych. Courtesy of Petro R. Sodol.

Ukrainian armed resistance to Soviet authority began sporadically with attacks against retreating Red Army troops, political prisons, and mobilization centers. In 1942, attacks on centers of the Nazi regime occurred haphazardly and armed resistance increased in the Volyn' Region and the swamplands of the Polissia Region. It was in this area that Taras "Bul'ba" Borovets organized a group of Ukrainians who fought against both Soviet partisans and German troops. In December of 1941, Borovets adopted the name *Ukrains'ka Povstancha Armia* (UPA) for his group.

Initially, both factions of the OUN were ill prepared to fight against the large, well-supplied Soviet and German armies. By early 1943, however, self-defense companies (OUN-SD) and OUN-M units began actively eliminating German outposts and battling marauding Soviet partisans. By July 1943, there was a recognized need to unite the three Ukrainian factions, but all attempts at dialogue failed. Over the course of the summer, OUN-SD units surrounded the other partisan groups and disarmed them. Many of the disarmed units were ultimately integrated into a new structure, the unified UPA. By the end of 1943, the UPA had effectively absorbed all other independent Ukrainian groups.

## UPA Organizational Structure

Figure 7. Taras "Bulba" Borovets, 1941.

In late November 1943, the UPA established a military structure and a Supreme Headquarters. Roman Shukhevych, also known by his military "nom de guerre" "Taras Chuprynka", rose to the rank of commander and Dmytro Hrycai assumed the duties of the Chief of Staff.

The areas of UPA operations were divided into four military zones, three of which were active: UPA-North, UPA-South and UPA-West. UPA-East remained inactive. Each of the military zones were further divided into Military Districts and in 1945 these were further split into Tactical Sectors. Each Tactical Sector had its own commander and six staff sections: operations, intelligence, logistics, personnel, training and political education. This structure remained until 1949.

On the battlefield, the UPA fielded companies. The size of the companies varied over time and place, but usually included 100-200 men. Companies were made up of three or four platoons, each of three or four squads. Between 1943-5, multiple companies often formed into battalions and on more than one occasion, two or three battalions united into combat brigades.

Figure 8. UPA Insurgents, 31 October 1944, from *Ukrains'ka Povstancha Armiia.*

## Bandera's Boys

The surviving documentation provides a good understanding of how each infantry squad function. Each squad consisted to ten to twelve men, armed with one or sometimes two light machine guns, two or three automatic weapons and the remaining men were armed with bolt action weapons. Specialized units existed as well, and these could be cavalry, mortar teams, reconnaissance troops, snipers, heavy machine guns, and artillery.

UPA had a daunting task of filling mid and low-level officer positions. Often Ukrainians who had served in the Soviet, Polish, or German armies were promoted but there were not enough people to fill these roles. Military veterans from the post WWI era were often too old to fulfill active combat roles, but a number of them served on staff organizations. To meet the needs of the UPA, each military district organized its own NCO schools for those insurgents that showed leadership skills. The UPA also ran two officer candidate schools, which graduated approximately 500 men in 1943 and 1944.

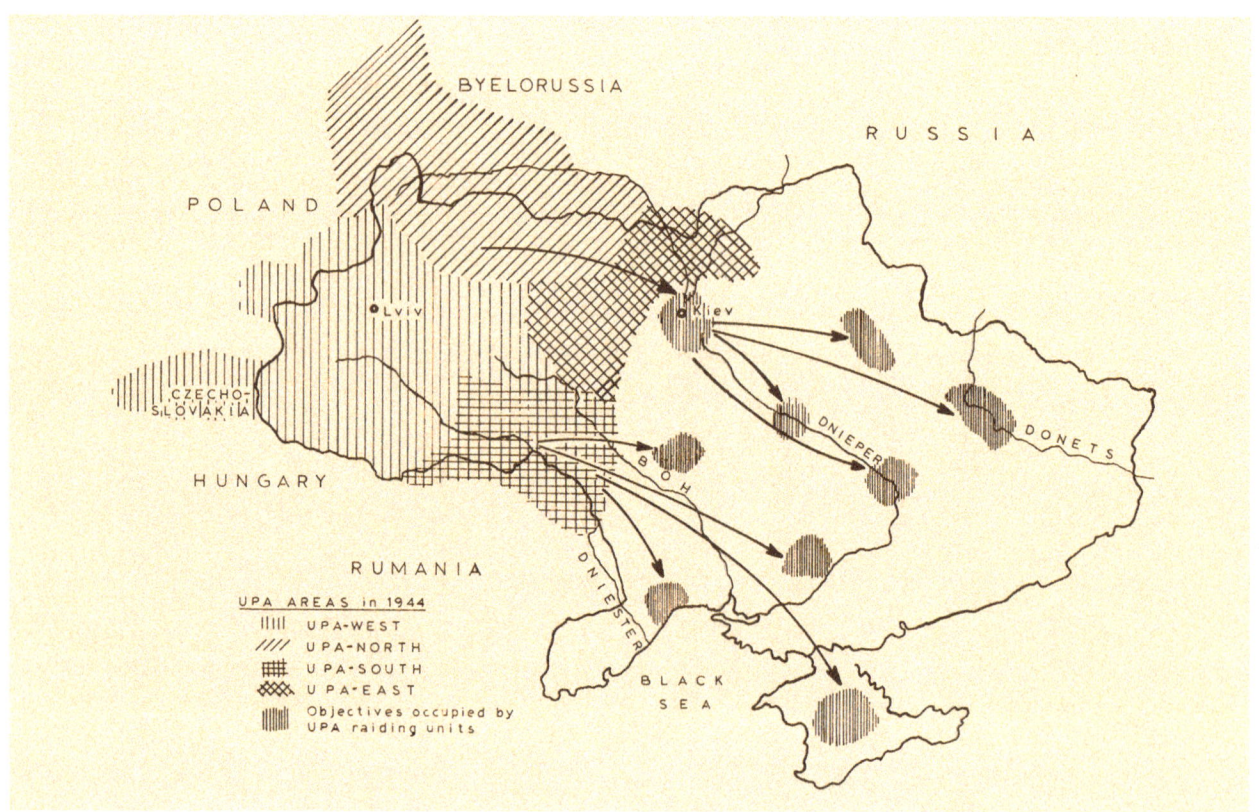

Figure 9. Map of UPA zones, from *UPA Warfare in Ukraine*.

By October 1944, the last units of the German army had been pushed out of Ukraine and the Soviet security forces launched their first major offensive against the UPA. By the spring of 1945, Soviet authority was limited to urban centers and the insurgents controlled much of the countryside. Soviet attempts to use returning Red Army units to combat the UPA proved to be ineffective, as the returning divisions were exhausted and often deserted in the face of the insurgence. The Russians assigned new NKVD troops and began to blockage the countryside. Each village was garrisoned by NKVD troops. Recognizing that the Soviets needed to break the bond between the local Ukrainian population and the UPA, the Russians began to field specialized units that attacked the Ukrainian farmers. By dressing up as members of the UPA, these specialized Russian units perpetrated numerous atrocities against the local population. By the winter of 1945/6, the insurgents were forced into hiding and by the spring, most UPA units were decimated. Without outside help, the UPA could not survive another winter and that year, most units demobilized. Over the next few years, the remaining combat units either sought to escape to the west or continued to launch raids against the Soviet authorities. In March 1950, the UPA commander Roman Shukhevych was killed in a skirmish outside of L'viv. Later that year, Vasyl Kuk, who had previously headed the OUN organization in the eastern

Figure 10. Roman Shukhevych, November 1943.

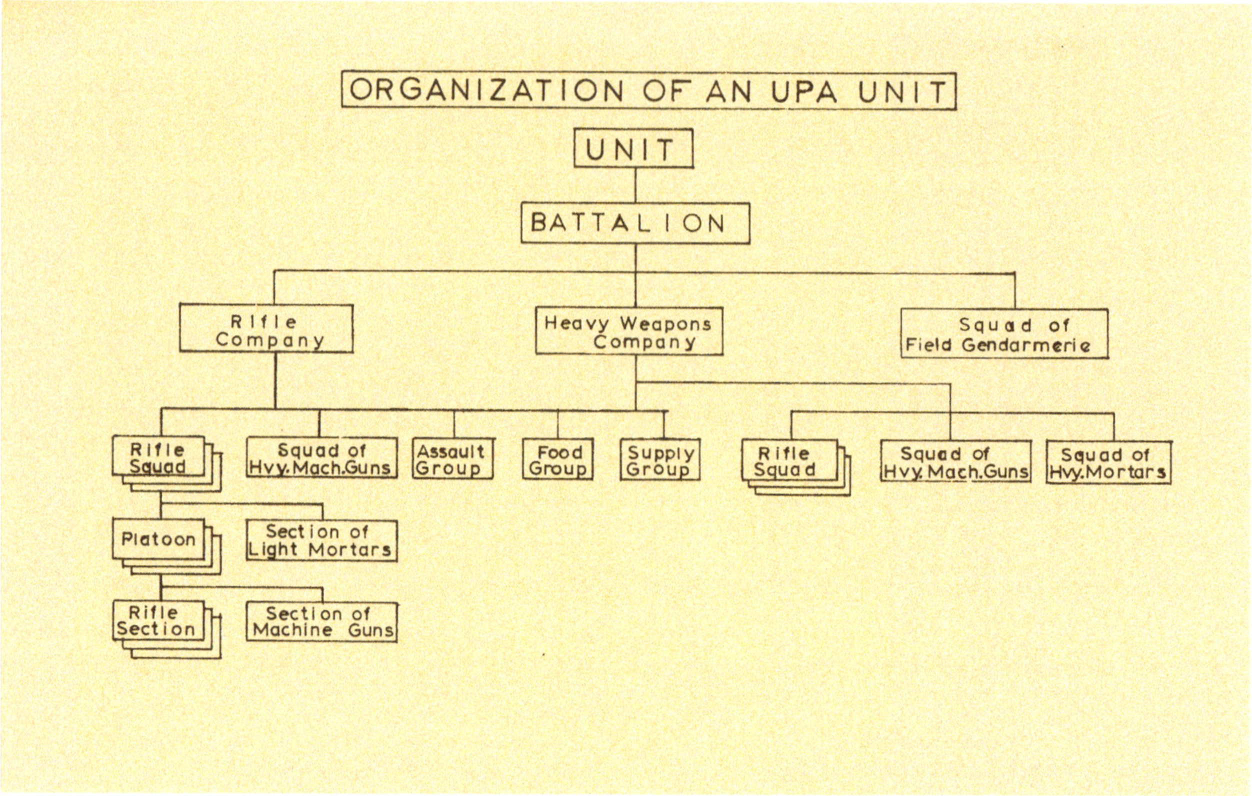

## ORGANIZATION OF AN UPA UNIT

```
UNIT
 │
BATTALION
 ├── Rifle Company
 │    ├── Rifle Squad
 │    │    └── Platoon
 │    │         └── Rifle Section
 │    └── Squad of Hvy. Mach. Guns
 │         ├── Section of Light Mortars
 │         └── Section of Machine Guns
 ├── Heavy Weapons Company
 │    ├── Assault Group
 │    ├── Food Group
 │    └── Supply Group
 └── Squad of Field Gendarmerie
      ├── Rifle Squad
      ├── Squad of Hvy. Mach. Guns
      └── Squad of Hvy. Mortars
```

Figure 11. UPA organization, from *UPA Warfare in Ukraine*.

Dnipropetrovsk region of Ukraine, became the leader of the underground struggle. Among the last reported raids by UPA members occurred in 1960, when two men and a woman launched an attack the town of Berezhany.

Figure 12. Vasyl Kuk.

# Scenario #1, Assassination of a High-Ranking Nazi, May 1943 - Klevan

*Overview*

In early May 1943, General Viktor Lutze, Commander of the Nazi SA (*Sturmabteilung*), was targeted for assassination. Like many in the Nazi party, Lutze viewed all Ukrainians as sub-human and stated on more than one occasion that all nationally or ethnically conscious Ukrainians needed to be eradicated. Known as a hangman of Ukrainians, members of UPA North began exploring ways of ending his activities.

Figure 13. General Viktor Lutze, Bundesarchiv. B 145 Bild - F051632-0523. Courtesy of Jurko Datsko.

*Introduction*

Learning from a spy that the general was planning on inspecting an area where a company strength UPA unit was active, the Ukrainian insurgents focused on set up an ambush to liquidate him. The UPA commander, *nome de guerre* "Vovchak," set up his ambush about three kilometers outside the town of Klevan. At this location, the road cut through a dense forest. In this area of Ukraine, dirt roads are often partially sunken into the ground, which allow the surrounding sides of the roads to act as banks along a streambed. Thus, those traveling along the road are confined with the sunken road and cannot easily move off to the side.

To lull the Nazis into a false sense of security, in the days preceding the attack, the UPA commander avoided the area of the forest. Thus, when the German patrols checked the forest, they were able to accurately report that there was no sign of any insurgent activity.

On departing the city of Rivne, General Lutze proclaimed that he would liquidate not only the UPA company but would also do the same to the local Ukrainian population. To do so, he brought with him a convoy of trucks, armored cars, and SD security troops on motorcycles. Lutze was also accompanied by an indeterminant number of Nazi high officials, including *Wehrmacht* and *Gestapo* officers, as well as members of his staff. These officials all travelled in cars.

To liquidate Lutze, Vovchak's four platoon company was strengthened with an additional machine gun section and two assault groups, both of which were armed with submachine guns and grenades. One platoon would be deployed to prevent the Germans from proceeding further along the road, while another one would be tasked with two tasks - closing the rear of the column and to prevent the Germans in the town from coming to the aid of their comrades. The heavy machine gun section was deployed along the slopes Hill 224 to provide fire support to the UPA below. Once the Nazi's were boxed in, the two remaining platoons would provide close fire suppression as the two assault groups would storm the cars of the German officials.

As any insurgent activity could be noticed by those Germans patrolling the area, Commander Vovchak was unable to set up any roadblocks or blockades before the ambush was sprung.

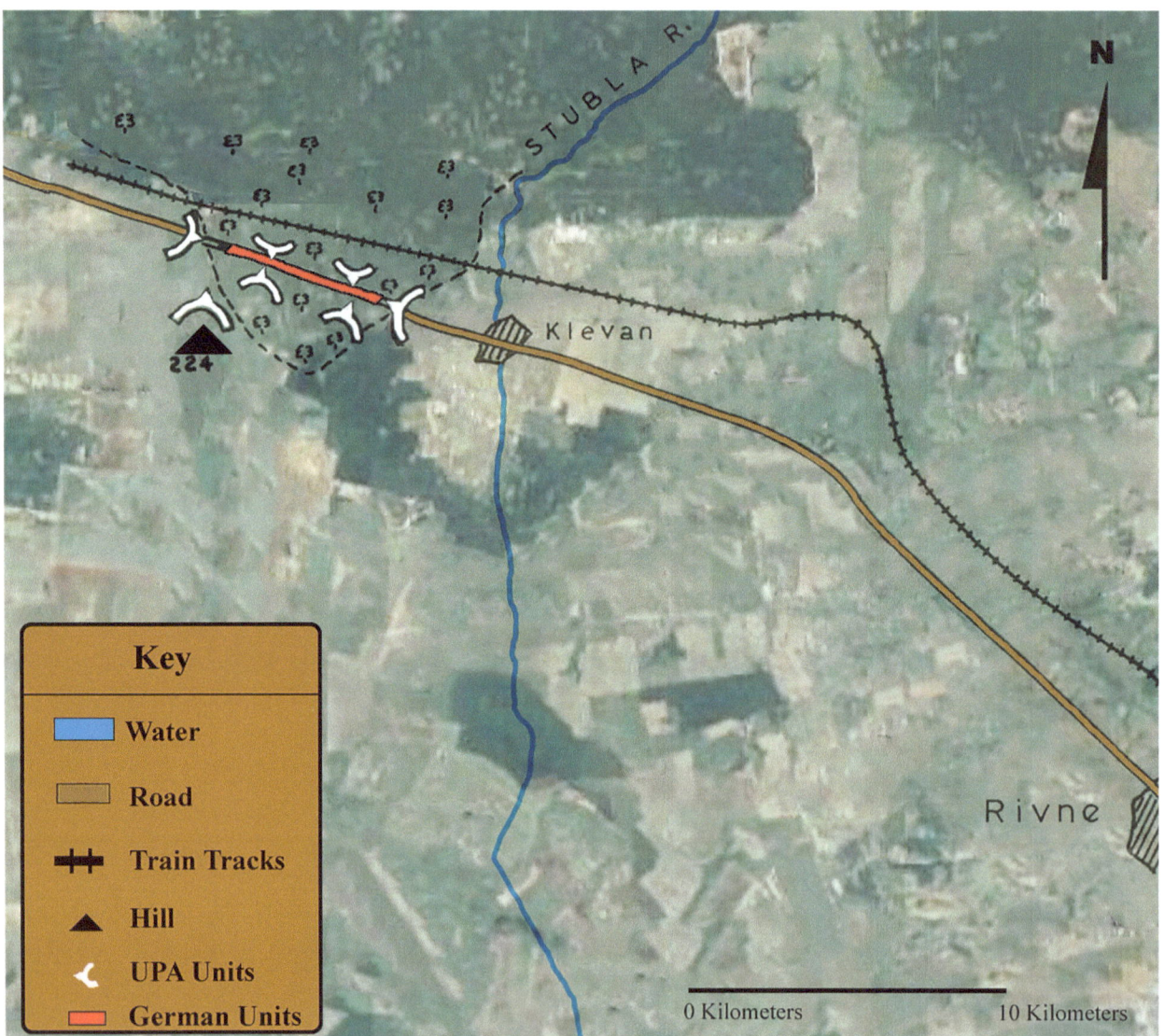

Figure 14. Scenario 1 Map.

## The Historical Engagement

The Nazi column departed Klevan at what was described as "fast clip". As the Germans entered the kill zone, the "easternly most" deployed platoon allowed the vehicles to pass by and only moved to block the road when the last Nazi had moved passed them. Once the cars had reached the area closest to the assault squads, the closest UPA squads began to fire. Seconds later the other UPA groups began to fire as well. Caught in the crossfire from both sides of the road, the German security troops and the officers sought to exit their vehicles and began to fire into the surrounding forest. The vehicles caught at the end of the column were unable to turn around and became pinned down by the heavy machine guns. Those in the front of the column also became pinned down by the fire from Krylaty's platoon.

The assault platoons quickly rushed toward the Nazi general and Lutze was riddled by a burst of machine gun fire. Once the UPA had successfully liquidated their primary target, Commander Vovchak ordered a retreat.

The UPA assignation remains poorly known. During the war, Nazi officials officially denied the assassination of their general and other high-ranking officials by the UPA. In its place, the Nazi's issued an official communique which reported that General Lutze was killed in an "auto accident". Parroting German sources, on 3 May 1943 the New York Times reported on page 8 that "Nazi Storm Troop Chief Badly Hurt in Accident". On 7 May Lutze received a lavish state funeral in the Reich Chancellery. Among the attendees were Adolf Hitler, who used the opportunity to speak about the dangers of driving at speeds over 50 miles per hour.

Figure 15. UPA Reenactor examining a German Sd.Kfz. 251 half-track, courtesy of Greywolves Co. Reenactment Group.

Bandera's Boys

# Order of Battle

## **Nazi**

Overall commander, General Lutze

Car #1. Passengers in the car include Lutze, an aide, a driver, and his blond secretary. Each is armed with a pistol and two figures are armed with SMGs.

Car #2 - four people (ranking officer, a driver, and two aids), four pistols, and two SMGs.

Car #3 - 3 people (ranking officer, an aid, and a driver), three pistols and one SMGs.

Truck #1 - Veteran infantry squad
NCO + 9 men, armed with 2 SMGs, 1 LMG, 7 rifles

Truck #2 – Regular Infantry squad
NCO + 9 men, armed with 2 SMGs, 1 LMG, 7 rifles

Truck #3 - Police Osttruppen squad
NCO + 9 men, armed with 2 SMGs, 1 LMG, 7 rifles

Truck #4 – driver and aide
Each man carries a pistol

SdKfz 222 Armored Car

Motorcycle troops #1
NCO + 9 men, armed with 2 SMGs, 1 LMG, 7 rifles
    Note: the LMG is mounted on the sidecar

Motorcycle troops #2
    NCO + 9 men, armed with 2 SMGs, 8 rifles

Figure 16. Bandai 1/48 German motorcycle and sidecar, author's collection.

## **UPA**
Overall commander - Commander Vovchak
Vovchak and 1 rifleman, armed with 1 SMG and 1 rifle

Platoon #1 under Oleksiy Shum
NCO + 19 men, 3 armed with SMGs, 1 LMG, 16 rifles

Platoon #2 under Krylaty
NCO + 19 men, 3 armed with SMGs, 1 LMG, 16 rifles

Platoon #3 under Kubik
NCO + 19 men, 3 armed with SMGs, 1 LMG, 16 rifles

Platoon #4 - Inexperienced
NCO + 19 men, 3 armed with SMGs, 16 rifles

Assault Platoon #1
NCO + 19 men, all armed with SMGs

Assault Platoon #2
NCO + 19 men, all armed with SMGs

Heavy Machine Gun Section containing
1 command group – 1 officer + assistant

4 x HMGs, each with three men
    Note - this group is located on Hill 224 and cannot move forward

Figure 17. 28mm UPA Figures, author's collection.

## Bandera's Boys

*Game Notes*

The UPA had chosen the perfect spot to set their ambush as the terrain had allowed them to place their assault squads within a turn's movement from the enemy. The high ground gave them an advantage when firing on the enemy below while the trees provided for the UPA. In addition, the nature of the sunken road confined vehicle movement.

For game purposes, only trucks or armored cars may try to move from the sunken road. If a truck or armored car tries to move off the road, there is a 50/50 chance they will be stuck. If a 1, 2, or 3 on a D6 (aka a six-sided die) is rolled, the vehicle in question becomes stuck trying to get out of the sunken road. To indicate a stuck vehicle, place the vehicle perpendicular to the road. Cars may not cross the embankment of the sunken road. To determine how motorcycle troops can move in the confined space, please consult your favorite rules.

Historically the cars travelled together as a group but for game purposes, the German player can set up his convoy in any manner that he chooses but the first two vehicles in the convoy may not contain Gen. Lutze. The UPA does not know which car is carrying General Lutze. Once the ambush is sprung, the German officers may leave their cars.

The game begins with the UPA having the first action.

Figure 18. UPA Insurgents from the group "Bohun" returning from a military action, 1943, from *Armia Bezsmertnykh*.

*Victory Conditions*

Most textbook ambushes focus on inflicting as many casualties on the enemy as possible, but in this action, all that matters is the execution of Lutze. The UPA win if Lutze is killed and the Nazi's win if he survives. Additional points may be awarded for the death of each Nazi officer or official while the Germans may gain points for the death of each insurgent. The game plays for 6 turns.

# Scenario #2, UPA attack on a German Military train, 23-24 June 1943 - Nemovychi

*Overview*

In the mid twentieth century, armies required substantial logistical support to remain active and the Nazi war machine was no different. Though Allied bombing sought to reduce these abilities by targeting key industries and production centers, rail lines remained the critical supply lifelines that allowed the Axis forces to function in the field.

The UPA, as has been noted previously, was self-sufficient and did not have the logistical support provided by that of a state. Although the local Ukrainian population provided food and intelligence to the insurgents, the small workshops scattered throughout the countryside were unable to manufacture medical supplies, weapons, or ammunition. Thus, the UPA needed to acquire all of their supplies from the enemy.

Though the UPA often stopped German and Hungarian trucks and cars for the purpose of capturing arms and ammunition, the spoils of war were usually limited to a few small arms and emergency medical first aid kits. Raids on towns, even if successful, were unpredictable as to what war trophies could be gained. Trains, on the other hand, could be an incredible goldmine of war material.

Figure 19. German troop train. Note *Flakvierling* quad AA.

## Introduction

In 1943, much of the Nazi war material for the eastern front went through Ukraine. Given the importance of trains to the German war effort, these mile long monstrosities were protected by a variety of means: armored engines, anti-aircraft train cars, specifically armored cars, and of course, guards. To reduce the possibility of being bombed from the air, Nazi trains also operated at night.

On the night of June 23/24 two UPA companies led by Commanders Dorosh and Yarema set their sights on stopping a train. The ambush was laid in a spot where the forest closed in on both sides of the track. The rails were smeared with soap to ensure the train was unable to stop before the detonating the mines placed by the insurgents. The two companies were deployed on both sides of the train track and occupied a 500-meter-long stretch of ground. Heavy machine guns supported the squads deployed along the front and rear of the ambush.

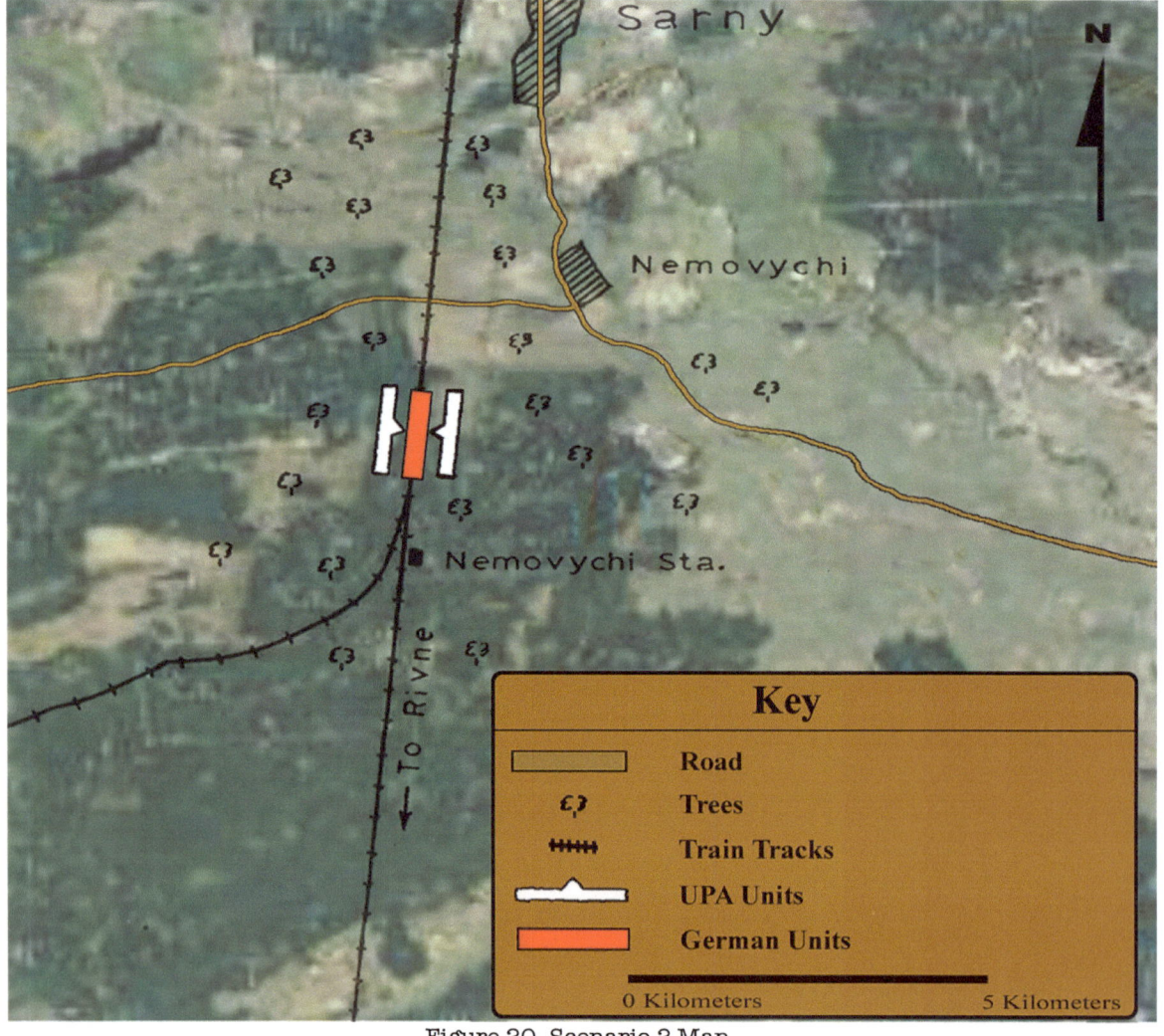

Figure 20. Scenario 2 Map.

## The Engagement

Sometime in the middle of the night a train approached the soaped-up section of track. Unable to speed up or slow down, the train slide forward until it hit the previously placed mines. Once the mines were detonated, the explosion wreaked havoc on the half-asleep Germans. Seized with panic, the semi-armed Nazis began jumping out of the train. As the Germans began to flee the train, they were fired upon by the insurgents. A few groups made it into the woods and began to fire on the UPA. At this point, the insurgents regrouped and were able to destroy the un-coordinated Nazi defenses.

With almost all of the Germans neutralized, the UPA began to take stock of their loot – and it was impressive. Among the trophies captured by the insurgents were heavy weapons, small arms ammunition, military field equipment, clothing, medicine, and food. Though both UPA Commanders Dorosh and Yarema were slightly wounded in the attack, the action was considered a great success.

Figure 21. Group of UPA Insurgents, from *Ukrains'ka Povstans'ka Armia*.

Bandera's Boys

# Order of Battle

**Nazi**
Overall commander, none

SD Security Police
Officer with SMG + 1 rifleman

4 full squads of infantry, each of 1 NCO and 9 men armed with 2 SMGs, 1 LMG and 7 rifles

1 MMG team

2 x 20mm Flak 38s mounted on the train

Figure 22. 28mm German miniatures, courtesy of Matthew Doyle.

**UPA**
Overall commanders, Dorosh and Yarema

Deployed along one side of the track
Commander Dorosh (SMG) + 1 rifleman

Platoon #1
1 NCO + 19 men, 3 armed with SMGs, 1 LMG, 16 rifles

Platoon #2
1 NCO + 19 men, 3 armed with SMGs, 1 LMG, 16 rifles

Platoon #3
1 NCO + 19 men, 3 armed with SMGs, 1 LMG, 16 rifles

Platoon #4 - Inexperienced
1 NCO + 19 men, 2 armed with SMGs, 18 rifles

1 x HMG three-man team

Deployed along the other side of the track
Commander Yarema (SMG) + 1 rifleman

Platoon #1
1 NCO + 19 men, 2 armed with SMGs, 2 LMG, 16 rifles

Platoon #2
1 NCO + 19 men, 2 armed with SMGs, 1 LMG, 17 rifles

Platoon #3
1 NCO + 19 men, 2 armed with SMGs, 1 LMG, 17 rifles

Platoon #4 - Inexperienced
1 NCO + 19 men, 2 armed with SMGs, 18 rifles

1 x HMG three-man team

Figure 23. 28mm UPA Assault Squad, primarily armed with SVTs, author's collection.

<u>Game Notes</u>
"Life is like a box of chocolates, you never know what you're going to get"

Forrest Gump, 1994


The goal of the ambush was to capture much needed supplies from the Germans. To see what is liberated from each car, a die roll is needed to determine what is found in each car and the value of the items. The number of points collected at the end of the game determines the level of victory or loss.

The existing literature is unclear if the UPA had specifically targeted this particular train or did they simply attack the first train that rolled into their ambush. If attacking the same train that the UPA attacked the night of 23-24 June 1943, follow the Nazi Order of Battle provided above.

The second option, based on the premise that this train happened to be in the wrong place at the wrong time, randomizes both the trains defenders and its cargo. As such, it can lead to a more unpredictable game. Such a scenario also allows for solitary play, as the one player first places his UPA troops on the table and then based on the random resulting defender rolls, places the Axis forces. All German ten men squads consist of 1 NCO and 9 men, armed with 2 SMGs, 1 LMG, and 7 rifles. Each train has two 20mm Flak 38 at the front and the rear of the train.

## Random Train Table (D10)

1. Military supply train with:
    Reinforced Platoon of SD Security Police
        Officer, 4 ten-man squads of infantry, 1 MMG team

2. Troop train with:
    *Wehrmacht* Soldiers
        Officer, medic, 4 ten-man squads of infantry, 2 MMG teams

3. Troop train with:
    *Wehrmacht* Soldiers
        Officer, 4 ten-man squads of infantry, 2 MMG teams
        In addition to the two 20mm Flak 38 at each end of the train, this train also has one 88mm Flak 36 Gun in the middle of the train

4. Troop train with:
    *Waffen* SS troops
        Officer, 3 ten-man squads of infantry, 2 MMG teams

5. Military supply train with:
    *Osttroppen*
        Officer, medic, 4 ten-man squads of infantry, 2 MMG teams

6. Military supply train with:
    8th SS Cavalrymen. The horses are in a separate wagon and play no part in the game. Officer, 4 ten-man squads of infantry, 2 MMG teams

7. Troop train with:
    Paratroopers
        Officer, 1 six-man squad of attached Pioneers with a flamethrower, 3 ten-man squads of *Fallschirmjager*, 2 MMG teams

8. Military supply train with:
    Hungarian troops
        Officer, medic, 4 ten-man squads of infantry, 2 MMG teams

9. Military supply train with:
    *Osttroppen*
        Officer, medic, 4 ten-man squads of infantry, 2 MMG teams

10. Military supply train with:
    SD Security Police
        Officer, sniper, 4 ten-man squads of infantry, 2 MMG teams

# Bandera's Boys

In addition to the engine, the train is pulling a number of cars. While historically some trains could stretch for miles, we have found that trains pulling more than 10 cars is difficult to duplicate on the table. Similarly, when gaming with six cars, the results are too unpredictable and the game can end before it truly begins.

Figure 24. UPA Reenactors, courtesy of Greywolves Co. Reenactment Group.

Once a UPA unit is within one turn's movement from reaching the car, the Nazi player rolls a D20 to determine what is in a particular train car. If the train car contains combat troops, the squad may activate at the start of the next turn. If one is using Warlord's *Bolt Action* rules, an additional die is added to the dice cup at the end of the turn. Once the UPA member reaches the train car and opens it, the German player needs to declare what the insurgents have captured.

*Freight Car (D20)*

| | | |
|---|---|---|
| 1 | Self-sealing stem Bolts | 0 points |
| 2 | Limited amount of 7.92mm rifle ammunition | 5 points |
| 3 | French artillery pieces, no ammunition | 0 points |
| 4 | Large caliber shells | 2 points |
| 5 | Advanced medical supplies | 8 points |
| 6 | Obsolete French rifles, some ammunition | 2 points |
| 7 | Anti-Tank and Anti-personnel Mines | 9 points |
| 8 | Scrap iron | 0 points |

| 9 | Squad of SS veterans (they will try to kill you) | 0 points |
| 10 | Tires, various sizes | 1 point |
| 11 | Grenades and 7.92mm rifle ammunition | 7 points |
| 12 | Vehicle parts | 1 point |
| 13 | Uniforms | 3 points |
| 14 | Horses | 10 points |
| 15 | 150 *Nebelwerfer* Shells | 6 points |
| 16 | Motorcycles | 8 points |
| 17 | Squad of *Wehrmacht* recruits | 0 points |
| 18 | Tins and Cans of Food Stuff (MREs) | 12 points |
| 19 | Mortars and ammunition | 16 points |
| 20 | Advanced and Prototype Nazi Weapons | 20 points |

The game begins after the mines have been successfully detonated. The train at a dead stop but is sitting upright on the track.

Figure 25. UPA Insurgents, from *Iavorivs'kyj Fotoarkhiv UPA*.

## *Victory Conditions*

Each dead combatant is worth 5 points. The side with the most points at the end of the game wins.

The game plays for six turns.

## Scenario #3, Liberation of a Penal Labor Work Camp, July 1943 - Sviatoslav

*Overview*

In the middle of the night, the Ukrainians successfully raid a Nazi slave labor work camp.

Figure 26. Area of the Slave Labor Work Camp, 2018, author's photograph.

*Introduction*

When the Germans invaded the Soviet Union in 1941, they intensified their use of slave labor in the recently captured areas. The reason for this was two-fold, to increase German economic exploitation and to speed up the mass extermination of non-Aryans. Among the Nazi run organizations was the Ukrainian Fatherland Service, which was based on the "Baudienst" building service, which was active in Poland.

The initial plan was to provide workers with housing in military barracks, limited food rations, and work clothes. By law all Poles and Ukrainians from the age of 14 to 60 were required to work for the Nazis. Many individuals were deported to Germany itself, where they worked in agriculture or factories. Those individuals who were troublesome to the Nazi authorities were incarcerated in special "Schtraflager" penal labor camps, such as the quarry at Sviatoslav, near the town of Skole.

Located in the foothills of the Carpathian Mountains, the camp conditions at Sviatoslav were designed to cause and inflict as much pain as possible before an individual died of exhaustion and malnutrition. Poorly fed and housed in a series of wooden barracks, the prisoners were required to mine rock from the nearby mountainside. The prisoners, who worked from sunup to sundown, were sent to work wearing only shorts and wooden shoes. Subject to daily beatings, the malnourished prisoners were dying in large numbers. According to one report, 200 young men were dying every month.

News of the camp conditions reached local OUN activists, who began to formulate a plan of action to rescue their compatriots. Observing the camp from the surrounding forests, it quickly became apparent that more than half of the camp inmates were in no condition to travel by themselves and would require transportation. The other complication was though the camp itself was rather isolated, a large German and Hungarian garrison was located only a few kilometers away in the town of Skole. Thus, any attack on the camp would require a quick and complete liquidation of the Gestapo garrison before they could summon reinforcements.

A local OUN activist, Mykhailo Lutsyk, came up with a plan to rescue the 500 Nazi prisoners. Using his local contacts, he arranged for up to sixty wagons to transport those prisoners, a former WWI medical officer to oversee to their immediate needs, food, and villagers to hide the prisoners during their recovery. To prevent the Nazis from easily recapturing or reinforcing the camp, two platoons were ordered to set up ambushes along the road from Skole. One platoon was dispatched just outside of Skole and the other closer to Sviatoslav. The third platoon was to rush the camp from all sides. Once the OUN members made it past the guard towers and wire fence, they were to concentrate on the Gestapo headquarters and police barracks.

Figure 27. UPA Insurgents, from *Iavorivs'kyj Fotoarkhiv UPA*.

## *The Historical Engagement*

Early in the evening, the first two platoons positioned themselves to delay any reinforcements reaching the camp. They took up positions that overlooked the road and if pressed by the enemy either directly or on a flank, they could escape to the rear into the mountains. Meanwhile, the third platoon prepared to launch their multi-pronged attack on the camp. Sometime after midnight, the Ukrainians launched a coordinated attack. Initially shooting from the tree line, the OUN members first neutralized the sentries manning the guard-towers with their bolt action rifles. Cutting through the electrified fence, the Ukrainians then rushed into the camp itself and were able to capture the MG outposts that were set up within the camp itself. So sudden and unexpected was the attack that only the guards on duty mounted any sort of defense. The troops in the barracks, as well as the majority of the Gestapo personnel, were captured still in their beds. Knowing that it was only a matter of time before the Nazis would be sending reinforcements, the OUN members quickly executed their Gestapo prisoners on the spot, gather as many supplies as possible, and began moving their prisoners away from the camp. The 250 prisoners who were unable to walk were loaded into the wagons. As it was unpractical to house all of the former prisoners in one mountain village, the men were separated into groups and were sent into the numerous villages a few days march from the camp. Once the last prisoners were removed from the camp, the two supporting platoons also disappeared into the neighboring forests.

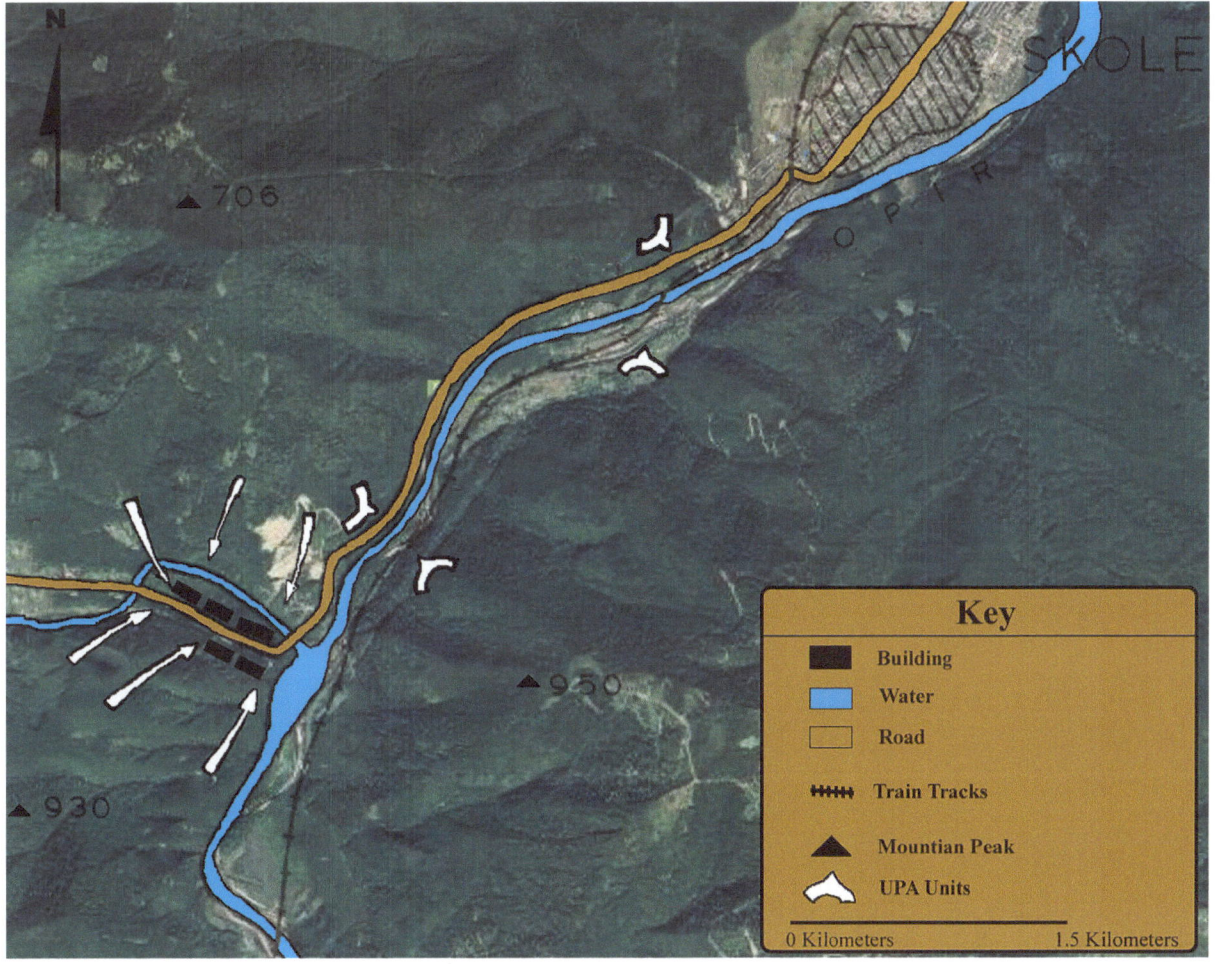

Figure 28. Scenario 3 Map.

The following morning, OUN operatives in Skole noted a big commotion. Both Hungarian and German troops were sent to the camp and finding nothing, began searching the neighboring villages. Unable to find any trace of the former prisoners, the Nazis took out their frustration on the local Ukrainian peasants. Villagers were tortured, raped, or simply executed on the spot, and three villages were burned to the ground.

# Reconstructed Order of Battle

## Nazis

Two men in each guard tower. They are armed with a rifle and a SMG.

Four men in each MG position. They are armed with one MG, one SMG and 4 pistols.

Twelve men in the *Gestapo* headquarters building. Each man is armed with a pistol and there are 3 SMGs and 4 rifles in the building.

Forty men are sleeping in the German police barracks. They are armed with 4 SMGs and 20 rifles in the barracks.

Figure 29. 28mm German camp personalities, courtesy of Warlord Games.

## OUN

Overall commander
Mykhailo Lutsyk, one medic, one additional rifleman

1st Platoon (Deployed guarding the road to Skole)
28 men, divided into three squads. They are armed with two SMGs and the rest are armed with rifles.

2nd Platoon (Deployed guarding the road to the camp)
40 men, divided into four squads. They are armed with three SMGS and the rest are armed with rifles.

3rd Platoon
Squad #1, 14 men
Commanded by Luka Toliubas. They are armed with one SMG and the rest are armed with bolt action rifles.

Squad #2
Commanded by Lev, 12 men. They are all armed with bolt action rifles.

Squad #3
Commanded by Dmytro Suslynets, 16 men
Suslynets is armed with the one SMG and the rest of his men are armed with bolt action rifles.

Figure 30. 28mm UPA figures, author's collection.

## Game Notes

One can recreate this action on the tabletop in one of two ways – either gaming the entire raid and include the possibility of Nazi reinforcements appearing in time to change the outcome of the assault or gaming the attack on the camp by the 3rd platoon. If one decides to game the smaller size attack, then one can also include the possibility of an errant enemy patrol showing up. Although one can use whatever favorite rules to game the raid, Warlord's *Bolt Action* lends itself directly to recreating the liberation of the camp.

To represent the camp, the game requires four buildings holding prisoners, one Gestapo headquarters, one camp guard barracks, three guard towers and three MG positions. As the OUN had observed the camp before the attack, the location of both the police barracks and the Gestapo headquarters should be known to the Ukrainians. The OUN should also be able to position their troops around the camp without the Nazis seeing them. Once the OUN troops begin shooting, any Germans on duty should be able to react. The MGs were positioned to fire upon the prisoners and not on the tree line, so they will need to be shifted before being able to fire on the attackers. The MGs are protected by sandbags.

Capturing enemy weapons is also considered to be a priority by the OUN but it is clearly secondary to the liberation of the prisoners. Any Ukrainian unit can use any recently captured small arms, but any shooting should be severely penalized (firing should be at a minus 2).

As the Ukrainians close in on the sleeping Gestapo (12 inches of the building), the German player rolls three D6. On a roll of a 6, the alarm is raised but the Germans will need a turn to equip themselves. For game purposes, the Nazis are considered to be disordered. If the German player fails to roll a six on the three D6s, the Ukrainians can try and rush the building without taking any defensive fire.

The prisoners themselves are inactive during the initial assault. Once a OUN squad reaches a building housing prisoners, the Ukrainians in that building can begin to move away from the fighting toward the waiting wagons. The prisoners move automatically on their own accord at the end of a turn. It takes two turns for the prisoners to reach the wagons and get into them. The following turn, the wagon moves away from the camp and is considered to have escaped.

Figure 31. 28mm Hungarian miniatures, courtesy of Warlord Games.

During the assault, there was always the possibility that an errant Nazi patrol would move toward the camp. At the end of turn Four, the Nazi player rolls a D6 and applies the result at the start of the following turn. The reinforcements enter at the eastern edge of the table.

**Die result**:

1 - no one shows up

2 - a 10-men squad of drunken German Wehrmacht decided that they needed to get some air. They are armed with rifles, pistols, and one SMG. Because of their condition they shoot at a minus 1.

3 - a 12-man Hungarian infantry squad is returning from leave. They are armed with rifles and two SMGs.

4 - Twelve *Gestapo* men are returning to the camp in an unarmored truck. They are all armed with SMGs and pistols.

5 – A twelve-man SS unit tasked with liquidating Jews is in the area and moves to the sound of the fighting. Armed with two MGs, 4 SMGs and rifles, they will focus on destroying the wagons and any prisoners.

6 – An opened top SdKfz 222 armored car appears on the road. It is armed with a MG and a 20mm cannon. It can move on the road and the flat area of the camp only.

<u>Victory Conditions</u>

The end goal for the OUN is to remove as many prisoners from the camp as possible. The more prisoners reach the wagons and disappear into the countryside, the greater the victory.

The game plays for 8 turns.

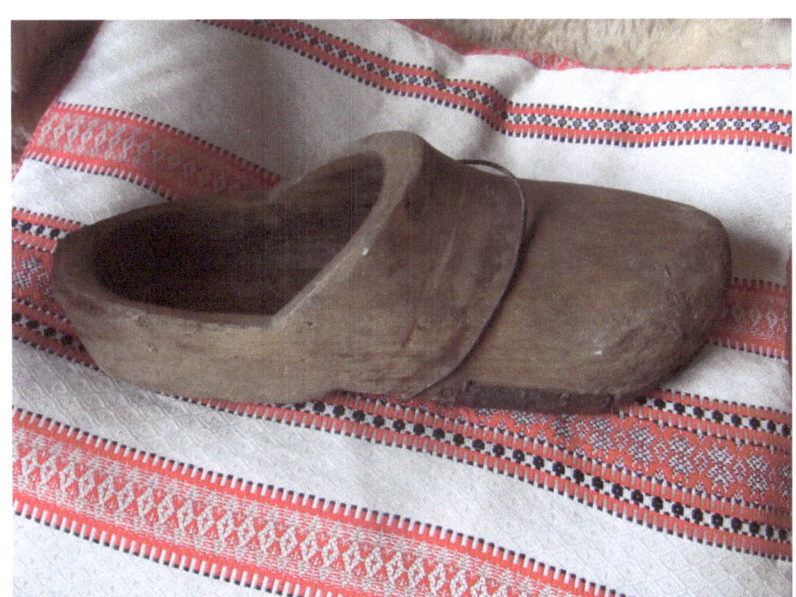

Figure 32. Wooden shoe worn by camp prisoner, 2018, author's photograph.

# Scenario #4, Raid on a town, August 1943 – Kamin Koshyrsky

*Overview*

Following the UPA assassination of General Lutze, the Nazis carried out a series of repressions against the civilian population. The village of Kortylisy, which unfortunately found itself near the spot of the ambush, was the target of a punitive operation. Civilians were executed on the spot and the village was burned to the ground. A group of women and children, who hid in the village church during the reprisals, suffered a particularly gruesome death when the building was torched by the Nazis.

Aware that the Germans were planning on unleashing additional acts of terror from their base at Kamin Koshyrsky, the UPA decided to launch a pre-emptive raid on the town. Before the war, the multi-ethnic and multi-religious town was home to several thousand citizens. Under the Nazis, the town's Jewish population was first confined to a ghetto and then exterminated. In their place, German troops were billeted in the settlement.

Figure 33. Pre-war photograph of Kamin Koshyrsky showing the town's central square, from volyntimes.com.ua.

## Introduction

The town is located along the banks of the Tsyr River. This north flowing body of water is between 1 to 3 meters deep and flows through the center of town. During the war, a number of bridges spanned the river. In the center of town, along the right bank of the river is a small mound, known as "The Castle." Though no longer standing, the area of "The Castle" contains the remains of a 12th century settlement. Monumental stone buildings, such as the 17th century Dominican monastery, were fortified by the Germans and new bunkers were built by the occupation.

The UPA plan for attacking the city called for a small group of men to infiltrate into the city and make their way to the Germans within the city center. Once the special "commando group" reached their objective, the insurgents would launch a three-pronged supporting attack. In the resulting chaos, the Germans would not know from which direction the enemy was attacking and would be unable to launch a successful counterattack.

The UPA forces available for the attack consisted of the Nazar-Kryka battalion, a company under *Lysyi* (the Bald), and a second company under *Kubik* (the Cube). The overall commander of the operation was *Rudyi* (the Red or the Ginger). For this operation, the insurgents were also able to call on a few mortars and a 75mm gun.

Commander *Rudyj* hand-picked 30 of the most experienced insurgents for the assault group, which was placed under the command of Commander *Chernyk*. The raiders were armed with four machineguns and the rest were equipped with submachineguns. The men also carried pistols, hand grenades, axes, saws, gasoline, and matches.

## The Historical Engagement

Shortly after dusk, the raiders reached the outside of the town. Removing their boots, the men crept toward the first houses of the settlement. All was quiet and the only sounds were those of the German sentries marching on guard duty. Moving quietly through the town, the men were able to avoid most of the German guards. One guard was less fortunate, as he was grabbed and quickly eliminated.

From here the group broke into three. The first group moved toward the open German positions, the second began dousing the barracks building with gasoline, and the third moved toward where the Germans were quartered.

As the attack proceeded, a group of the raiders noticed a room in the German barracks with a light. Believing it to be a guard house, three UPA men threw open the door and found all the Germans were in bed asleep, with only an interpreter and a member of the staff awake. With a quiet yet firm whisper of "Hands Up", nine Germans were taken captive.

Further down the road, gunshots were fired as the barracks building began to smolder and then burn. Sensing they were being attacked, the troops in the headquarters of the SD security police began firing blindly into the streets. Within minutes, the sleepy town was transformed into a battleground. Seeing the attack was now in full swing, the raiders fired off their red flares, which was to signal the remaining UPA to launch their attack.

As the UPA raiders waited to be relieved, the Germans began to organize themselves. Small groups of Germans began to encircle the base of "the Castle" where the raiders were concentrated. Some fool-hearty Nazis began to advance, but their initial poorly planned attacks were beaten off. As more German troops arrived, the raiders began to wonder when their comrades would arrive.

In what seemed like a lapse of 25 minutes, those on the outside of the town launched their attack. Two large explosions rocked the town and then they were repeated. The Germans who had encircled the base of "the Castle" retreated and those who remained in the buildings surrendered to the UPA. By 5:00 the town was in the hands of the insurgents.

Figure 34. Scenario 4 Map.

Figure 35. UPA insurgents in various scales, courtesy of Mykyta Karpukhin, https:www.facebook.com/karp.mykytin?mibextid+LQQJ4d.

During the raid, the UPA seized 20,000 rounds of ammunition, five machine-guns, four machine pistols, over 100 pistols, four radio receivers, eleven horses, seven motorcycles, quantities of food stuffs, and uniforms. The German police garrison was destroyed, 200 *Wehrmacht* personnel were captured, along with an undisclosed number of German fliers from the neighboring German airbase. Those Germans identified as taking part of the punitive raid on the village of Kortylisy were executed by the UPA on the spot. During the operation, the UPA reported that two men were wounded.

Figure 36. 28mm German Aircrew, painted by Michael Adams, courtesy of Jon Russell.

Bandera's Boys
# Order of Battle

## Nazi
240 policemen, armed with rifles and pistols
200 *Wehrmacht* soldiers, armed with machineguns, submachineguns, and rifles
12 *Luftwaffe* airmen, armed with pistols

## UPA
Senior Officer: *Rudyi* (the Red or the Ginger)

Assault Group
Commander *Chernyk* and his assistant, both armed with SMGs, flares, and grenades

Squad #1
10 men, armed with 1 machinegun, 8 SMGs, 10 pistols, grenades, and gasoline

Squad #2
10 men, armed with 1 machinegun, 8 SMGs, 10 pistols, grenades, and gasoline

Squad #3
10 men, armed with 2 machineguns, 6 SMGs, 10 pistols, grenades, and gasoline

Reinforcements
Nazar-Kryka battalion

Independent company under *Lysyi* (the Bald)

Second Independent company under *Kubik* (the Cube)

75mm gun

Mortar company

Figure 37. Unknown UPA Insurgent, from *Ukrains'ka Povstancha Armiia*.

**Game Notes**

There are a number of ways of wargaming the UPA raid on the town – the entire operation, the initial assault, or the relief operation. In any such tabletop recreation, however, the uneven number of combatants would require the game mechanics be adjusted in order to give both sides an opportunity to claim victory.

Another option is to begin the game with the UPA capturing a number of buildings and awaiting reinforcements. The Germans are scattered, but with each turn they can consolidate their available forces or try and rush the insurgents with the troops currently available. If playing such an action, each turn the German player would role to see what troops become operational and where they would appear on the table. Such a game is well suited to Warlord's Bolt Action rule set.

To determine Nazi reinforcements, at the start of each turn the German player rolls three D20 on the Reinforcement Chart to see what troops become available.

# Bandera's Boys

**Reinforcement Chart**

1. no troops appear
2. 6 unarmed policemen
3. 4 *Luftwaffe* pilots, armed with pistols
4. 6-man Regular *Wehrmacht* Squad, armed with 1 SMG and 5 rifles
5. Inexperienced Medium Machine Gun Team
6. 6-man Inexperienced *Osttruppen* squad, armed with 1 SMG, and 5 rifles
7. Regular sniper team (can appear only once. If a second 7 is rolled in the game, only 2 Nazi units appear on the table for that turn).
8. 5-man Inexperienced *Osttruppen* squad, with 1 SMG, 1 MG, and 3 rifles
9. Inexperienced Light Mortar team
10. German Medic and 2 stretcher-bearers, unarmed
11. Regular SdKfz 251 with pintle-mounted MMG and 6 men armed with rifles (can appear only once. If a second 11 is rolled during the game, an Inexperienced SdKfz 222 appears on the table. No more than two German vehicles can appear on the table).
12. Regular Flamethrower team (can appear only once. If a second 12 is rolled during the game, only 2 Nazi units appear on the table for that turn).
13. Regular *Wehrmacht* 2$^{nd}$ Lt and a veteran soldier, both armed with pistols
14. Veteran Medium Machine Gun Team
15. 6 policemen, all armed with rifles
16. 6-man Regular *Wehrmacht* Squad, armed with 1 SMG and 5 rifles
17. 5-man Veteran *Wehrmacht* Squad, armed with 1 SMG, 1 MG, and 3 rifles
18. Veteran *Wehrmacht* Major and a veteran soldier, both armed with SMGs(can appear only once. If a second 18 is rolled during the game, only 2 Nazi units appear on the table for that turn).
19. 6-man Veteran *Wehrmacht* Squad, armed with 1 SMG, 1 MG, and 4 rifles
20. 10-man Veteran *Wehrmacht* Squad, armed with 1 SMG, 1 MG, and 8 rifles. The German player then rolls a D10 to see where each unit appear off the game table. The placed troops must be placed within 6 inches of the table edge.

*German Placement*

1. UPA player can pick which location the German troops appear
2. Northeast corner of the table
3. Norther edge of the table
4. Northwest corner of the table
5. Western edge of the table
6. Southwest corner of the table
7. Southern edge of the table
8. Southeastern corner of the table
9. Eastern edge of the table
10. German player can pick which location the German troops appear

The figures are then placed on the table. If using Warlord's *Bolt Action* rule set, now place an additional order die for each new unit into the dice cup. The German player is free to use these units at the start of the turn.

To keep game play less predictable, after Turn 4 the UPA rolls a D6 to see when their reinforcements start their artillery bombardment which would in essence end the game. Thus, on a roll of a 1 or a 2, the UPA reinforcements would begin their attack the following turn. On a roll or a 3 or a 4, the UPA reinforcements would start their attack in two turns; a roll of a 5 or a 6 would require the UPA assault squad to hold on for three more turns.

## *Victory Conditions*

The UPA needs to hold the two buildings and the trench they are in. The Nazi player needs to dislodge the insurgents. If the UPA hold all three objectives at the end of the game, they gain a major victory and if they hold two objectives, they have a minor victory. If the Germans hold all three objectives, they gain a major victory and if they hold two objectives, they have a minor victory.

Figure 38. UPA reenactors, courtesy of Greywolves Co. Reenactment Group.

# Scenario #5, German attack on a UPA NCO school, fall 1943 – Black Forest

## *Overview*

By mid 1943, the tide of war was being turned against the Germans. Beginning in January US bombers flying from England were hitting targets in Germany and by May the Nazis were driven from Africa. On the Eastern Front, the Nazi's failed to capture Stalingrad and were being driven westward. Lacking any substantial reserves, the Germans sought to establish a defensive line along the Dnipro River. As it was critical for the Nazis to secure their supply lines, the Germans sought to clear the rear areas of Ukrainian insurgents and Soviet partisans.

For the UPA, the Nazi reversal presented a possible opportunity. Painfully aware that the lack of a standing Ukrainian army prevented the establishment of an independent Ukrainian state in the aftermath of WWI, the current leadership sought to expand its insurgency. Much effort was placed on collecting war material and storing it for the upcoming struggle. Recognizing that while many of the recruits were devoted to the oncoming armed struggle, the organization lacked trained leaders. To rectify this situation, the UPA established a series of NCO and officer schools in secure areas.

Figure 39. UPA NCO school taking oath, from *Ukrains'ka Povstancha Armiia*.

## *Introduction*

The Carpathian Mountains, which run through Romania, Hungary, Slovakia and Poland, are ideal for asymmetric warfare. The mountainous region, with its large Ukrainian population, supported the Ukrainian armed struggle for statehood. The rugged mountainous terrain, along with the region's lack of roads and multitudes of forests, all favored the insurgents. Over the last two years, the German and Hungarians had focused on garrisoning the larger towns, while the smaller settlements were only under their nominal control. By mid 1943, the UPA maintained numerous bases, hospitals, warehouses and training camps in the Carpathians.

To liquidate these UPA training facilities, from August to December 1943, the Germans launched a series of operations against the Ukrainians. Martial law was declared and military tribunals were instituted. It is estimated that the Nazis executed approximately 15,000 Ukrainians, either in public or in secret. Not all those who were killed were part of the underground movement, as the Germans sought to use terror to maintain order.

Between October and December, the Nazis attacked the four UPA military training camps set up in the Carpathians. Initial German reconnaissance provoked several skirmishes with the UPA. Recognizing that they were up against a well-organized fighting force, the Germans troops established a blockage of the area. Next, the Nazis concentrated a great quantity of light and heavy weapons, including artillery. Air support for their operations was provided by the Luftwaffe. In addition to deploying eight to ten police regiments, the Nazis also brought in SS troops. Lacking the personnel to attack the entire area of UPA activity simultaneously, the region was divided into four districts. Once an operation was concluded in one section, the troops were redeployed to attack the following district.

Figure 40. UPA reenactors, courtesy of Greywolves Co. Reenactment Group.

### The Engagement

On the morning of 27 November, the Nazis attacked a section of the Black Forest (*Chornyj Lis*). Following an aerial bombardment by the Luftwaffe, the Germans unleased an intensive artillery barrage on the area. German troops which were previously deployed in the villages of Maidan, Posich, and Zaviy, began to advance onto the camp that was located between the three villages. By noon, though a number of Nazi troops became disoriented in the mountainous terrain, the Germans were able to surround the camp by noon. Knowledge of the terrain not only allowed the UPA troops to avoid contact with the majority of the German troops but also gave them the opportunity to launch well-placed counterattacks.

In the process of breaking out of their encirclement, the UPA managed to capture the headquarters of the German units and their radio station. Deprived of their command and control, the Nazis were unable to proceed with their attack. At this point, the UPA troops were able to penetrate the German line in two places and then attacked the Nazis from the rear. As the Germans called off their attacks and sought to extradite themselves from the forest, the UPA units pressed their attacks on the German rear. The battle turned into a route and the Black Forest remained under UPA control.

Though the Germans were able to launch attacks on all four of the camps during their offensives, and in some instances burned structures and seized UPA supplies, the Nazis failed in their strategic goals of neutralizing the UPA units. The combat losses on the attack on the Black Forest remain unknown, but during the Nazi attack on the fourth camp Capt. Brandenburg reported a loss of 316 German police troops and that the insurgents suffered 34 men killed in action.

*Order of Battle*
See Game Notes below

*Game Notes – The Full Attack*
There are two ways of replicating the German attack on the camp. If one were to focus on gaming the entire attack, one would be best served by using a company level game, where a stand of figures represents is an infantry squad or a single vehicle. There are a number of rulesets that fit within these parameters, including Arty Conliffe's *Crossfire*. One advantage of *Crossfire* system is that the emphasis is on infantry combat. Another popular option is Battlefront's *Flames of War*. As each ruleset usually includes an organization chart tailored to their particular game system, players should rely upon a German Order of Battle included in their ruleset. If no organizational chart is provided, a generic German one is provided below.

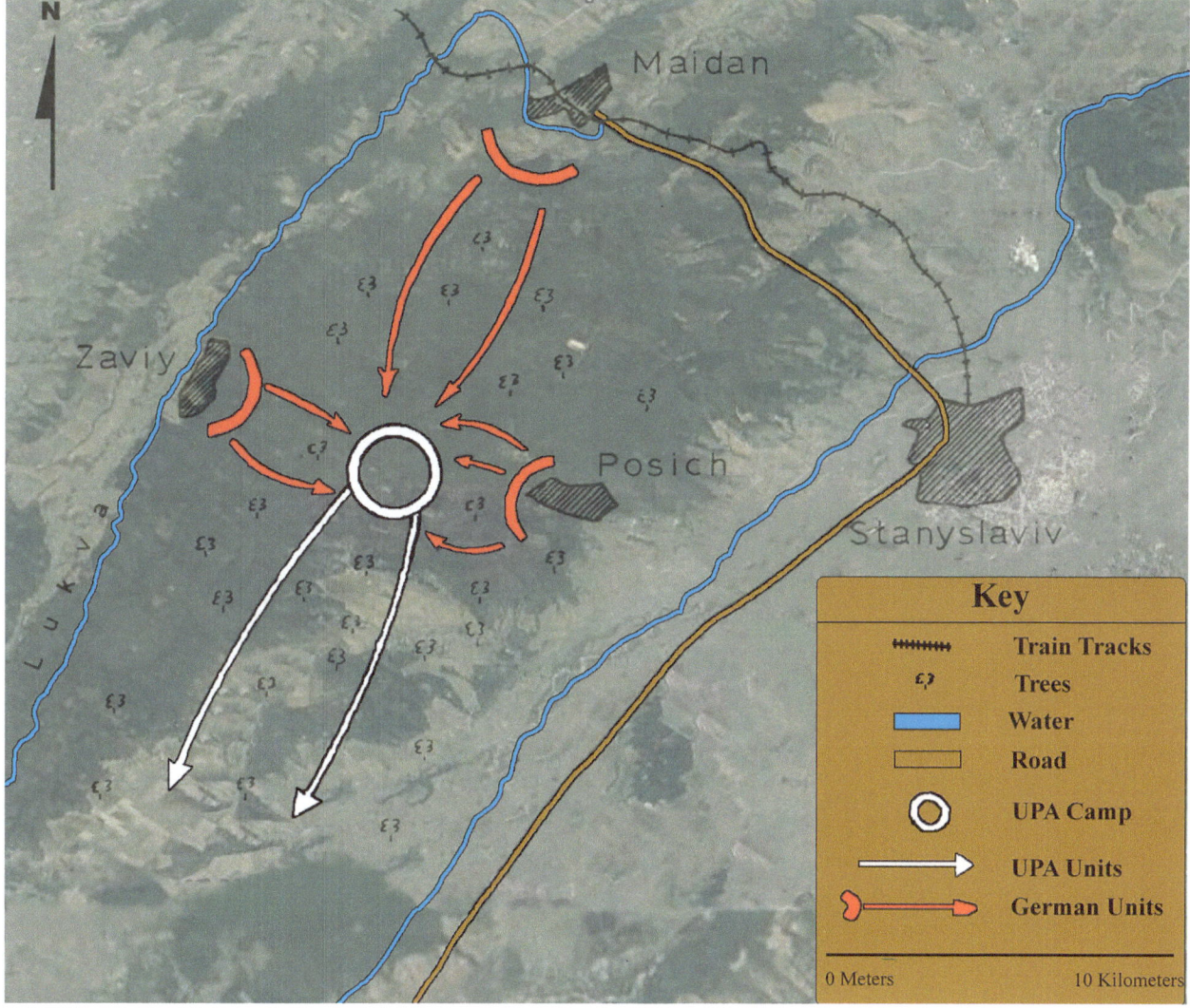

Figure 41. Scenario 5 map.

**Nazi**
Senior headquarters
HQ platoon

*SS Troops*
1 battalion of
 5 companies of infantry
1 pioneer company
1 machinegun company

*Police Troops*
3 regiments of Police Troops, each of

1 Regimental HQ platoon
3 battalions each of
    3 infantry companies (12 MGs and 3 50mm Mortars)
    1 support company (12 MGs and 6 81mm Mortars)
    1 cavalry platoon
*Artillery*
3 howitzer companies

*Air support*

## UPA
The UPA have an estimated 500 men in the Black Forest. They are divided into:
Senior command
Veteran unit
20 Recruit companies, armed with rifles and grenades
12 MG companies
4 mortar companies
1 *Nebelwerfer* company
4 snipers
12 mines

Figure 42. 20mm German troops, courtesy of Casey Pittman.

    To replicate the general chaotic nature of the Nazi operation, the scenario lends itself to have multiple German players (3 to 6 players) attempting to capture the camp. In such a scenario, the German players compete against each other. Point are awarded to the troops who manage to break into the camp first and capture the flag of the Ukrainians which is flying in the center of the camp. Additional points are awarded for capturing supplies and burning buildings, while points are deducted for the loss of German casualties.

Figure 43. 28mm UPA sniper team, author's collection.

There is a fine line when playing the UPA in this action. Historically, the insurgents completely outmaneuvered their attackers and this needs to be reflected on the tabletop. UPA's familiarity with the terrain should give them the ability to move freely, while the direction and rate of German movement left to chance. If your rules do not include such a possibility of random movement, the following table provides a quick and easy way of duplicating the chaos of mountain fighting. Before a unit moves, roll a D10 and apply the following results.

**Random Movement Table (D10)**

1 – turn the unit 90 to the right and move a full movement

2 – turn the unit 90 to the right and move a half movement

3 – no movement

4 – half move

5 or 6 – move as normal

7 – half move

8 – no movement

9 – turn the unit 90 to the left and move a half movement

0 – turn the unit 90 to the left and move a full movement

The UPA force is significantly outnumbered. Except for the teaching staff, the insurgent troops are all raw semi-trained recruits. They know how to fire their weapons but for most of them this is their first taste of combat. They also should act as small units while the Germans can fight in larger groups.

<u>Victory Points</u>

10 points - Capture the UPA flag in the center of the table.

5 points – capture one of three barracks buildings scattered within 12 inches of the flag

3 points – for each Nazi unit destroyed

2 points – for each UPA unit destroyed

Figure 44. Members of the UPA "Oleni" Officer Candidate School during their Oath of Allegiance ceremony, 21 September 1944, from *UPA: They Fought Hitler and Stalin*.

*Game Notes – Capture the German Headquarters/Radio Station*

The second option is to recreate one of the many small-scale skirmishes, including the attack on the German headquarters. For those using Warlord's *Bolt Action*, you can use the following list to organize your forces.

## **Order of Battle**

**Nazi**

Headquarters Group
Major +2 riflemen

Captain + 2 riflemen

Medic
Medic + 2 assistants

Squad #1, Veteran
10 men, armed with 2 SMGs, 1 LMG and 7 rifles

Squad #2, Regular
10 men, armed with 2 SMGs, 1 LMG and 7 rifles

Squad #3, Regular
9 men, armed with 2 SMGs, 1 LMG and 6 rifles

Squad #4, Inexperienced
8 men, armed with 2 SMGs, 1 LMG and 5 rifles

MMG Team, Regular
armed with MG42

Figure 45. 28mm German Troops, courtesy of Matthew Doyle.

## **UPA**
Command Group
Captain + 1 Rifleman

Squad #1, Veteran
10 men, all armed with SMGs

Squad #2, Regular
12 men, armed with 4 SMGs, 1 LMG and 7 rifles

Squad #3, Inexperienced
16 men, armed with 2 SMGs, 1 LMG and 13 rifles

Squad #4, Inexperienced
18 men, armed with 2 SMGs, 1 LMG and 15 rifles

Sniper Team, Veteran

Mortar Team, Regular

MMG Team, Raw

Figure 46. 1/144 German Henschel Hs 126 Reconnaissance aircraft, author's collection.

## *Victory Conditions*

If playing the larger game, the player with the most victory points wins. This game plays for 10 turns.

If playing the smaller skirmish, the UPA wins if they can capture or destroy the radio station. Anything else is a German victory. The game plays for 6 turns.

# Scenario #6, Hold the Line, April 1944 - Hurby

*Overview*

During the spring of 1944, the center for UPA-South was located in the Kremianets Forest, north of the city of Ternopil. The forest was also the immediate rear area of the 1st Ukrainian Front. Commanded by Red Army Marshal Heorhii Zhukov. Zhukov had recently replaced General Nykolai Vatutin, who was mortally wounded in a clash with the UPA back in February 1944. The continued combat operations of the UPA against the Soviet forces resulted in the NKVD launching a special operation to clear the forest area of all "enemies of the state".

Figure 47. UPA memorial at Hurby, 2011, author's photograph.

*Introduction*

On 21 April 1944, 30,000 Soviet troops were deployed to encircle the entire area of UPA activity. Under the command of General Marchenko, the 16th, 17th, 18th, 23rd, and 24th Rifle Brigades of NKVD Internal Troops began to push forward. Approximately 5,000 UPA soldiers, along with a large number of civilians, were caught in the trap. As the Soviet troops began to move forward, UPA units along the periphery sought to break free. By 23 April, following a number of heavy engagements, the main body of UPA troops under Brigade Commander *Yasen* was surrounded in the vicinity of Hurby. The UPA had well prepared positions and were supported with heavy machineguns, 120mm mortars, and light artillery.

At 0400 hours on 24 April, the Soviets launched their attack. In addition to the NKVD infantry units, the Soviets also committed artillery, tanks, and airplanes to the attack. The UPA defensive line held late into the night, but it was clear that the UPA would not be able to hold the line for another day. That evening, a number of UPA units were able to break out of the encirclement and fought their way through fresh NKVD blocking units. When the Soviets were finally able to secure the forest, they summarily executed any wounded UPA fighter or civilians that were unable to flee the area in time.

Soviet reports document that they were able to capture one operational aircraft, 7 artillery pieces, 15 mortars, 47 heavy machineguns, a printing press, 120 wagons, and significant quantities of food supplies. UPA sources estimate that they killed over 800 NKVD soldiers during the operation.

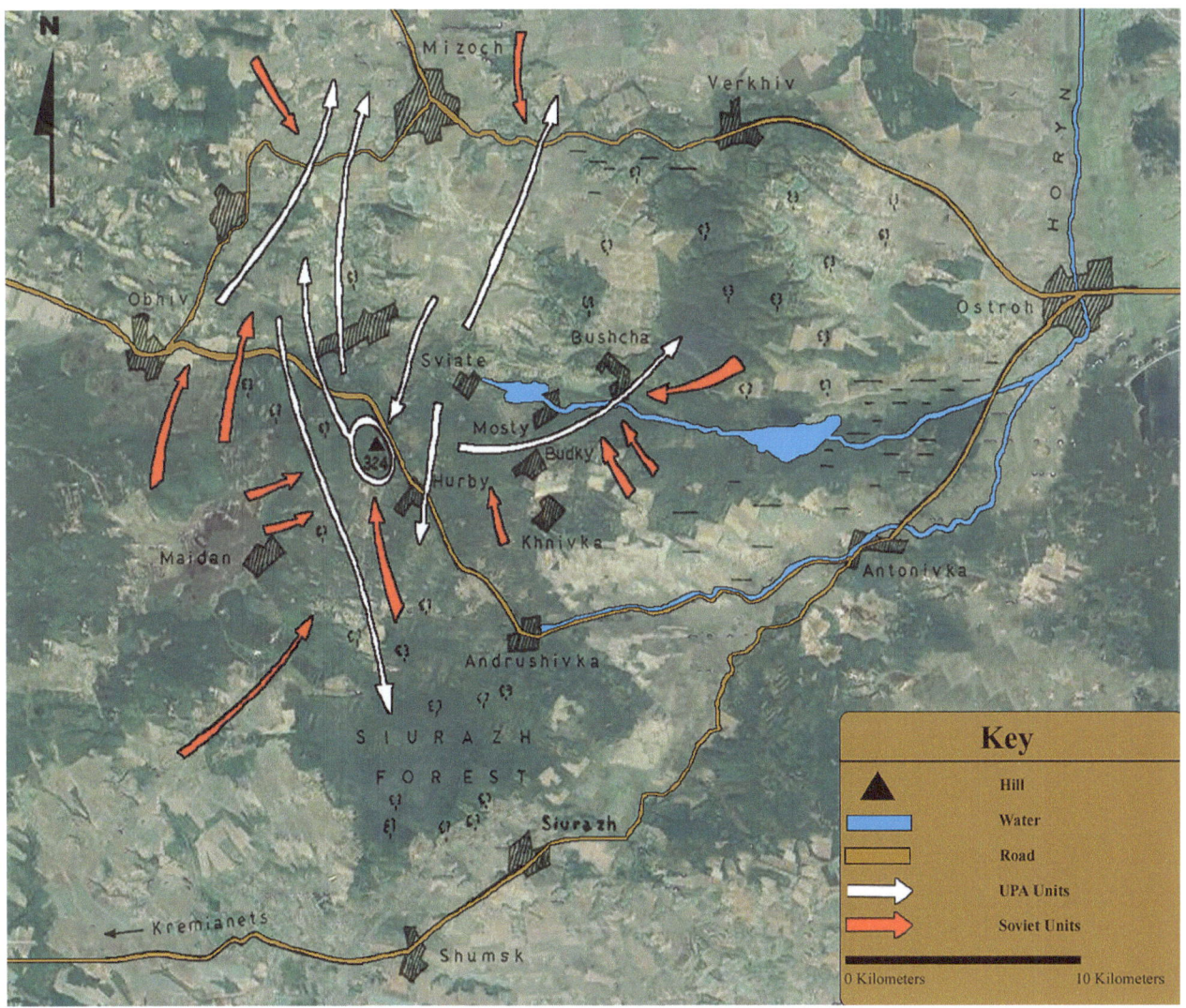

Figure 48. Scenario 6 Map.

## The Engagement

NKVD records show that 26 individual battles took place during the operation, including two on 21$^{st}$, three on the 22$^{nd}$, nine on the 23$^{rd}$, four on the 24$^{th}$, four on the 25$^{th}$, three on the 26$^{th}$, and two on the 27$^{th}$. This number does not include the countless skirmishes and smaller assaults that occurred that April.

## Game Notes

Rather than trying to replicate the entire operation, the scenario presented below focuses on one of the many desperate UPA actions of trying to hold the line. At their disposal, the UPA have acquired a number of weapons and ammunition from the retreating Axis, including a heavy mortar and an artillery piece. The defenders are well dug in and have two bunkers on the table. The area is sandy and tanks can get bogged when then get off of the roads.

The UPA can lay down four anti-tank mines before the start of the engagement outside the Soviet player's deployment zone. In our games we use twelve rocks, four of which have a red dot of paint on the bottom. When the NKVD unit moved to within three inches of a rock, it is flipped over. If no red dot appears on the rock, proceed as normal. If a red dot is revealed, the vehicle can no longer move but it can fire as normal the following turn. If the unit is not a vehicle, the unit which triggers the mine takes 3D6 damage and one D6 pin markers. Once detonated, the mine marker is removed from play.

The following Order of Battle works well with Warlord's *Bolt Action* rules.

Figure 49. 28mm Soviet SU-76 Self-propelled gun, courtesy of Warlord Games.

Adrian Mandzy

## Order of Battle

**Soviet**

Major + 2 men

Veteran Forward Observer

Regular SU-76

Regular BA-64

Veteran NKVD Company - All troops in this company are classed as *Fanatics*
Commissar + 2 men

2 Veteran NKVD rifle squads, each with
10 men, 3 SMGs, 1 LMG, and 6 riflemen

2 Regular NKVD SMG squads, each with
10 men, 1 LMG and 9 SMGs

Regular Light Machine Gun Squad
9 men, with 1 SMG, 2 LMGs and 6 riflemen

Inexperienced Flamethrower Team

Veteran Assault Engineer Squad

Figure 50. 28mm UPA Insurgents with heavy weapons, author's collection.

## UPA

Captain (Regular) + one man

Inexperienced Medic +1 man

2 squads of Veteran Guerrilla Fighters, each of
16 men, armed with: 4 SMGs, 2 LMG, and 10 rifles

1 squad of Regular Insurgents Squad of:
14 men, armed with: 3 SMGs, 1 LMG, and 10 rifles

2 Inexperienced squads of Insurgents, each of
14 men, armed with: 2 SMGs, 1 LMG, and 11 rifles

1 Regular Field Artillery

1 Regular Mortar Team

1 Regular Anti-Tank Rifle Team

1 Inexperienced *Nebelwerfer* or 1 Inexperienced 20mm *Flakvierling* 38

1 Regular Sniper Team

1 Inexperienced Machinegun team or Mortar team

Four anti-tank mines (see discussion above)

Figure 51. UPA reenactors, courtesy of Greywolves Co. Reenactment Group.

## Victory Conditions

The Soviet player needs to have at least one unit break into the rear of the UPA defenses. UPA units need to hold the line. The game lasts for 6 turns.

# Scenario #7, Winter Mountain Ambush, January 1945 - Kosmach

*Overview*

Within the Carpathian Mountains, the village of Kosmach and its surrounding hamlets served as one of the many supply and administrative centers for the insurgents. In January 1945, a force of Soviet NKVD security troops was sent in to eliminate Kosmach. Following an artillery barrage, the Soviets moved in to capture the settlement. The Soviets did not anticipate that the insurgents would put up such a determined defense and despite their numerical superiority, they were unable to take the territory. As the battle wore on, the forces of Moscow wired for reinforcements and an additional unit was sent to help to liquidate the Ukrainians.

Figure 52. Acting UPA commander Myroslav Symchych at the battlefield, 2 July 2012, author's photograph.

*Introduction*

Knowing that the Soviets had request additional troops, a young 22-year-old Myroslav Symchych, who served as the commander of the Birch Company, was tasked with keeping the reinforcements from reaching Komach. Knowing the terrain, Symchych set up an ambush a few kilometers outside of town on two ridges which overlooked the dirt road. At the spot where the road was cut by a small mountain stream, the insurgents set up their ambush. As they had arrived before the NKVD, the Ukrainians took apart the small wooden bridge that crossed the mountain stream. A short time later, when the twelve troop filled trucks and one car came upon the missing bridge, the vehicles stopped. The troops exited their transport and formed into two columns before beginning their advance on the village.

*The Engagement*

At this point Symchych gave the command to fire. Like wheat before a sickle, withering fire from the two ridges that overlooked the mountain road mowed down the invaders. Heavy fire was directed at both ends of the column and the enemy had no chance of breaking out of the ambush. Pinned between the two ridges, the enemy never recovered from the initial hail of fire. Symchych continued to direct the ambush until his company completely liquidated the enemy battalion. Hit in the arm as he moved between the troops, the UPA commander continued to give orders until the Soviet troops stopped firing. Weak from loss of blood, Symchych collapsed and was carried from the field. With the UPA leader out of the fight, no one gave the order to move off the ridges and collect weapons and supplies from the defeated enemy. Among those killed was the Soviet commanding officer Major General Nicholas Dergachov, who was best known for his punitive actions against Chechian and Tatar civilians.

In this textbook ambush, the Soviet forces were utterly destroyed. Following the initial salvo which ended the life of General Dergachov, the NKVD troops were unsure what to do and died where they stood.

Figure 53. Western ridgeline overlooking the ambush site, 2012, author's photograph.

## *Game Notes*

For game purposes, the amount of troops had been reduced. In place of twelve trucks, this scenario only uses four. Likewise, the number of troops has been reduced to make game play more manageable.

The UPA troops on the north side of ambush are halfway up on an unscalable mountain cliff face. Once the troops are placed in this location, they cannot advance toward the enemy. The UPA troops on the southern side of the ambush are a bit further away from the road and the tree covered slope is not a major impediment to movement. Historically, the UPA placed at least two machine guns, their anti-tank rifle team, and their mortar at the head of the column and two other medium machineguns to cover the back of the column, but in this game the UPA player can deploy his troops as they see fit.

The death of General Dergachov should play an important role in this scenario. At the start of the scenario, all Soviet troops can move as usual but once the General becomes a casualty, a die roll is required for each Soviet unit to test if they are able to move away from the kill zone. Inexperienced units need to roll a 6 on a D6, Regular units a 5 or a 6, and Experienced units a 4, 5, or 6. If the unit fails to make the required die roll on a turn, they gain a pin marker and cannot move that turn. The unit which failed a movement die role may try again on the following turn.

The Soviet player places his troops in the middle of the table. The NKVD troops, including the general, are all out of their vehicles and are lined up into two columns.

If using Warlord's *Bolt Action* the UPA player draws the first order die.

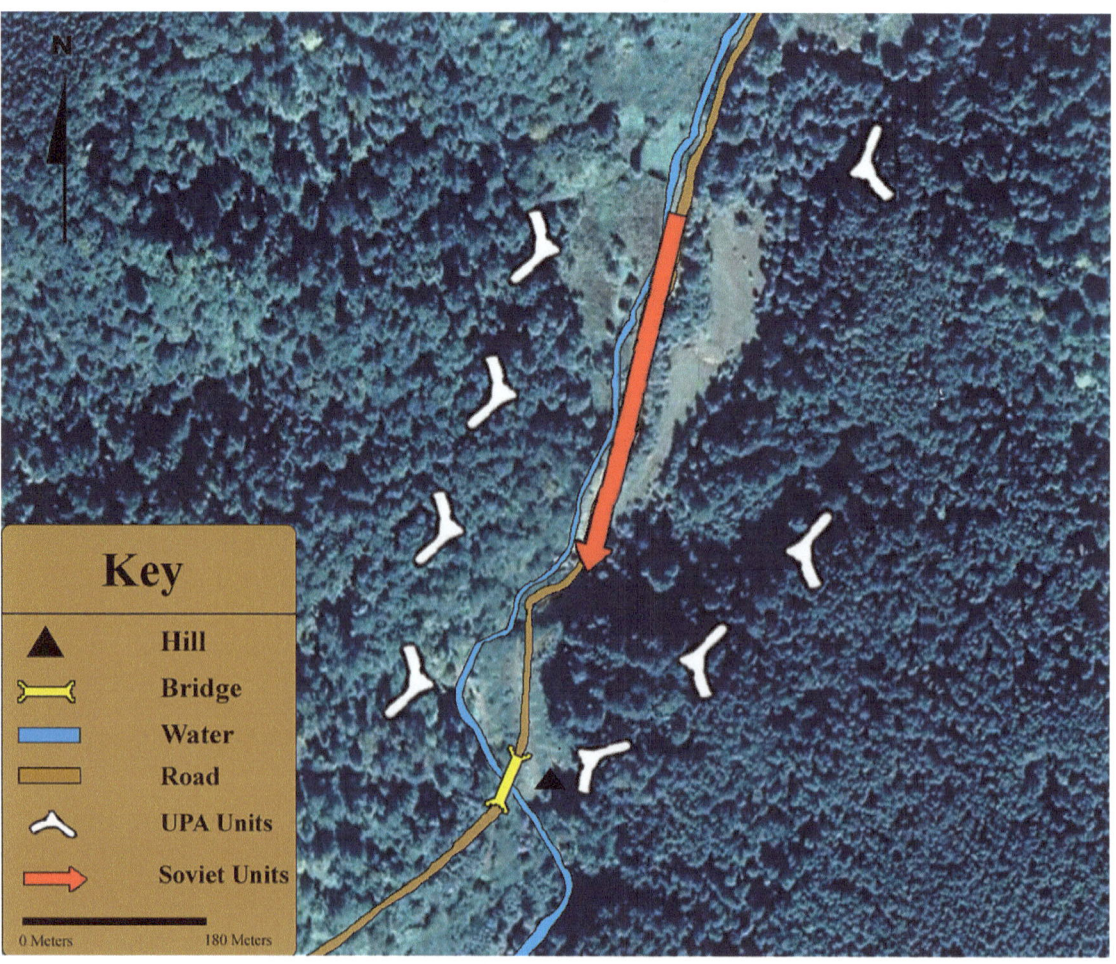

Figure 54. Scenario 7 Map.

## Order of Battle

**Soviet**

General + 1 aid

Medic + 1 assistant

2 Veteran NKVD rifle squads, each with
Ten men, 3 SMGs, 1 LMG, and 6 riflemen

2 Regular NKVD SMG squads, each with
Ten men, 1 LMG and 9 SMGs

4 Studebaker lend-lease trucks
1 Civilian command car

Figure 55. 28mm Soviet troops, courtesy of Warlord Games.

**UPA**

Commander + 1 rifleman

1 squad of Veteran Guerrilla Fighters of
16 men, armed with: 4 SMGs, 2 LMG, and 10 rifles

2 squads of Regular Partisan Squads, each of
14 men, armed with: 3 SMGs, 1 LMG, and 10 rifles

1 Inexperienced squad of Partisans of
14 men, armed with: 2 SMGs, 1 LMG, and 11 rifles

3 MMG teams, one Inexperienced, one Regular, and one Experienced

1 Experienced Anti-Tank Rifle Team

1 Regular Mortar team

1 Experienced Sniper

Figure 56. 28mm UPA troops in winter camo, author's collection.

## *Victory Conditions*

The Soviet player needs to get his troops off the table. For each NKVD figure moved off the table, the Soviet player gains one point. The General is worth 10 points if he gets off the table. Each vehicle that makes if off the table provides an additional 5 points to the Soviets.

The UPA player gains 10 points for killing the General and one point for each additional Soviet figure.

The game plays for 6 Turns.

Figure 57. UPA Insurgents, Birch Company, 1944, photo courtesy of Jurko Datsko.

# Scenario #8, Clash between UPA insurgents and Soviet partisans, February 1945 - Black Ridge

*Overview*

Following the Nazi invasion of the USSR in 1941, the Ministry of State Security called for Soviet partisans to be active in Ukraine. Few party members answered the call and the three largest groups were liquidated by the Nazis by the end of 1941. In June of 1942, the Ukrainian Staff of the Partisan movement was set up in Voroshylovhrad (modern day Luhansk). Under its director Nikita Khrushchev, Soviet partisans conducted an extensive raid from Russia into northern Ukraine. As the area was sparsely settled, the raid encountered little resistance from either the local population or the Nazi forces in the area.

Concerned with the growth of the UPA and wanting to demonstrate Soviet power, a second raid was launched in May 1943 under the command of Gen. Sydir Kovpak. This raid sought to cross the Volhynian Provence and establish a Red Army base in the Carpathian Mountains. In early August this group was crushed by the Germans at Deliatyn and its surviving bands were liquidated by the UPA. Additional Soviet paratroopers dropped into the area to assist Gen. Kovpak were disarmed by Ukrainian insurgents. Those partisans who survived the raid were assigned to the First Ukrainian Partisan Division under Petr Vershigora. As Soviet forces advanced through Ukraine, the Soviet partisan movement grew in number.

Figure 58. 28mm Soviet partisans, courtesy of Warlord Games.

The partisans set up an ambush near the Black Bridge, which crossed the Nezhukhivka River. At this point, the forest came close to the road, which provided the UPA insurgents with additional concealment. As the UPA company began to inspect the terrain before setting up their ambush, a group of UPA scouts that were following the Soviet partisans reported the enemy quickly approaching. Quickly disbursing, the insurgents set up make-shift positions and waited.

The UPA members did not have long to wait, as they heard the Soviets moving forward. The communists were all riding horses and were pulling wagons of plundered loot behind them. Discipline with the partisan ranks was loose. The men were riding by twos and fours. Some of the men were chatting and others were singing. One man was playing an accordion. Others were smoking cigarettes.

Figure 59. UPA Insurgent Mounted Reconnaissance Unit, UPA-North, Winter 1943-4, from *UPA: They Fought Hitler and Stalin.*

## *The Engagement*

When the Soviets were half-way across the bridge, the UPA commander gave the order to open fire. As the first group of Soviet partisans was hit, riderless horses began running – some toward the village from which they came, while others ran in the other direction. In the confusion, the partisans began firing. One of the Soviet machine guns jammed, but others kept on firing. Some of the partisans attempted to break through the ambush, while others dismounted and returned fire.

After a few minutes, the firefight was over. As the UPA fighters moved in to claim what supplies they could, they found the wounded Soviet commander playing dead. A quick prod proved that the officer was only simulating a corpse. The quick talking Soviet officer claimed to be a comrade, but an UPA insurgent replied that he was not a friend but rather a hangman and a thief.

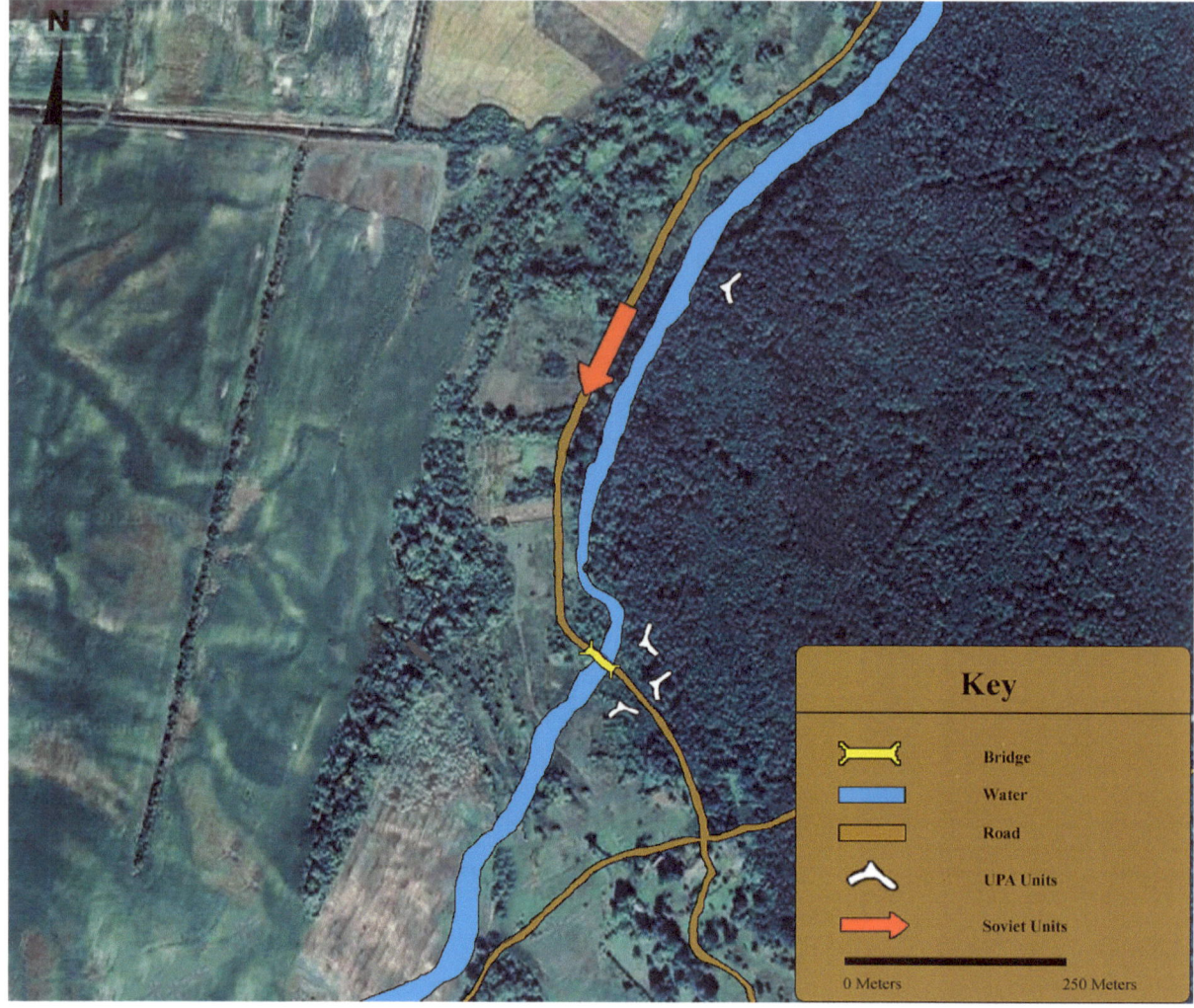

Figure 60. Scenario 8 Map.

## Order of Battle

*Soviet Partisans*
Commander
One Inexperienced Senior Lt and 2 Riflemen

Mounted Squad #1, Regular
12 men, NCO armed with SMG, 1 LMG, 10 carbines

Mounted Squad #2, Regular
10 men, NCO armed with SMG, 1 LMG, 8 carbines

Mounted Squad #3, Inexperienced

10 men, NCO armed with SMG, 9 carbines

Mounted Squad #4, Inexperienced
10 men, NCO armed with SMG, 9 carbines

Wagon #1, Regular
Three men armed with rifles

Wagon #2, Inexperienced
Three men, one with an SMG and two armed with rifles

Wagon #3, Inexperienced, Shirkers
Two men armed with rifles

Figure 61. 28mm UPA insurgents, author's collection.

*UPA Insurgents*
Commander
One Veteran Lt and 1 Rifleman, both armed with SMGs
Squad #1
10 men, NCO armed with SMG, 1 LMG, 8 riflemen

Squad #2
10 men, NCO armed with SMG, 2 additional SMG, 1 LMG, 6 riflemen

Squad #3
10 men, NCO armed with SMG, 1 LMG, 8 riflemen
Scouts

5 men, all armed with SMGs

Medium Machine Gun Team

*Game Notes*

Given the nature of this make-shift ambush, the UPA player is unable to prepare any fortified positions or deploy any mines. The river is fordable for all infantry and cavalry.

The UPA player, does however, get the first action of the first turn. If using Warlord's Bolt Action system, the first order die drawn for the scenario is for the UPA. If a Soviet order die is drawn before the first UPA order die, the Soviet player needs to move one of his units forward. If using Warlord's *Bolt Action* system, treat the mounted Soviet partisans as Cavalry. The fourth Soviet partisan squad and the third wagon crew should be treated as Shirkers as per *Bolt Action*.

Figure 62. UPA reenactors, Slavs'k, Ukraine, 2018.

*Victory Conditions*

The UPA needs to gain control of the wagons. The Soviet player needs to keep the wagons. The side that holds all three wagons at the end of 6 turns wins. Anything else is a draw.

Adrian Mandzy
# Scenario #9, Breakthrough into West Germany - 1947

*Overview*

In 1944, Stalin dictated new borders between the USSR and the Moscow installed Polish post-war government. As a result, civilians on both sides of the new border were subject to deportations and repatriations. Poles who had lived for generations in Ukraine were forcibly moved to areas which until recently were considered ethnic German territories. Ukrainians who found themselves now living in post-war Poland were also uprooted and moved. UPA units operating in these areas often intervened on behalf of the Ukrainian villagers being deported. Young people faced with deportation often joined the UPA at a time when other UPA units were being reduced or demobilized.

In the spring of 1947, the Polish government instituted Operation Wisla, which sought to remove by force the remaining ethnic Ukrainian population from the border areas and settle them in the former German lands now controlled by Warsaw. The deportations were accompanied by considerable violence, resulted in the deaths of some deportees while others suspected of aiding the UPA were imprisoned. The UPA sought to upend this process and attacked Polish units involved with the operation. At times, the UPA cooperated with Polish anti-communist groups, even taking part in joint military operations. In the end, however, with the Ukrainian population removed from the area, the civilian OUN network was neutralized and the UPA was deprived of their support network.

The UPA could not survive without receiving significant support from the local Ukrainian population. The previously friendly area they once defended was now filled with enemy troops and food had become scarce. With no one left to defend, the UPA had no reason to remain in the region. As UPA casualties increased, the remaining units were ordered to either cross into the USSR or to march across Czechoslovakia into West Germany.

Figure 63. UPA insurgents from the "Udarnyk-2" Company and US troops in West Germany, August 1947, from *Armia Bezsmertnykh*.

## Introduction

The UPA had conducted raids from the very inception. Though raids often have a military objective, the UPA also conducted raids as a form of informing people of their goals in creating an independent Ukrainian state. Even before the UPA's formal creation, its political leadership sent groups out to propagate the idea of Ukrainian statehood. As the conflict intensified, raids were also conducted into non-Ukrainian areas for the purpose of creating a network of supporters. Often such raids would break through a border area and spend weeks organizing mass meetings with villagers where they would explain what the UPA is and what they were fighting for. The UPA leadership spent a good deal of its resources in creating pamphlets, leaflets, and other printed materials for distribution during such raids.

The UPA had conducted many successful raids into Czechoslovakia, but by 1947 the Moscow installed Czechoslovakian government had sent troops to guard the mountain passes. As more UPA units sought to move toward the West, more Czechoslovakian troops were deployed. Despite the omnipresence of enemy troops and moving through unfamiliar territory, a number of UPA units were successfully able to cross into Austria and West Germany for a period of two years.

The UPA raids had a long-lasting impact. Not only did it allow a number of UPA combatants a way to physically survive, but their presence gave a voice to the ongoing Ukrainian national liberalization struggle in the Ukrainian diaspora communities. According to the UPA leadership, the raids to the west were to inform the world of the UPA struggle and to demonstrate that a military resistance to Moscow continued. Given that the UPA members continued to reach the US Zone of Occupation over the course of two years, their presence caused the Soviet authorities to generate a plethora of reports to explain away the UPA phenomenon. No longer in a position to simply deny the existence of the UPA, Moscow embarked on a campaign of dis-information that strove to link the Bandera movement as "Nazi" – a trend that continues to this day.

Figure 64. Living History Timeline display of Weapons and Uniforms used by Ukrainian Forces, Kholodnyj Iar, Ukraine, 2017, courtesy of Jurko Datsko.

*The Engagement*

In May 1947, the UPA leadership sent three separate groups toward Bavaria. The first group, led by Hromeko, moved through the Great Tatra and Little Tatra Mountains, Prerav, Trebich, and Pilsen. On 11 August 1947, the remnants of the group reached Bavaria. During their travels, this group had marched over 1,500 kilometers and fought twenty-two separate engagements.

The Burlaka group, which initially consisted of three companies, left Ukrainian territory on 3 June 1947 and followed the path blazed by the Hromeko through the Great Tatra and Little Tatra Mountains. Shortly thereafter, the Burlaka group was encircled by the Vah River near Zilina by a force of 15,000 Czech troops. The Czechs forces, though were well equipped with armored cars, artillery, and reconnaissance planes, were unable to stop the UPA insurgents. In the ensuing engagement, the UPA fighters were able to break through and continued westward.

On 13 August, the Burlaka group found itself surrounded once again. At this point, the group was down to sixty-seven men. To avoid their total destruction, Burlaka broke his command down into seven smaller units and ordered each squad to break out in different directions.

Three weeks later, the remainder of Burlaka's squad was trapped once again by the Czechs. Once again Burlaka broke his men into two groups and were able to successfully break out in two directions. A few days later, Burlaka and his handful of followers took refuge in a forest ranger's hut. As the Czechs continued their pursuit, Burlaka found himself trapped once again by overwhelming forces. Unable to escape, he and another captured UPA insurgent was brought to trial in Prague. The Czechs sought to use Burlaka's image to call on other UPA insurgents to surrender. Held in a prison near the city of Koshyts. On 12 February 1948, Burlaka and five other insurgents broke out of prison but were recaptured a few days later.

Other UPA raiding groups continued to follow. In addition to the larger number of fighters moving west, the UPA also sent couriers through Czechoslovakia. Some of the couriers travelled west and other travelled back home to Ukraine.

Figure 65. 28mm NKVD/MVD figures, author's collection.

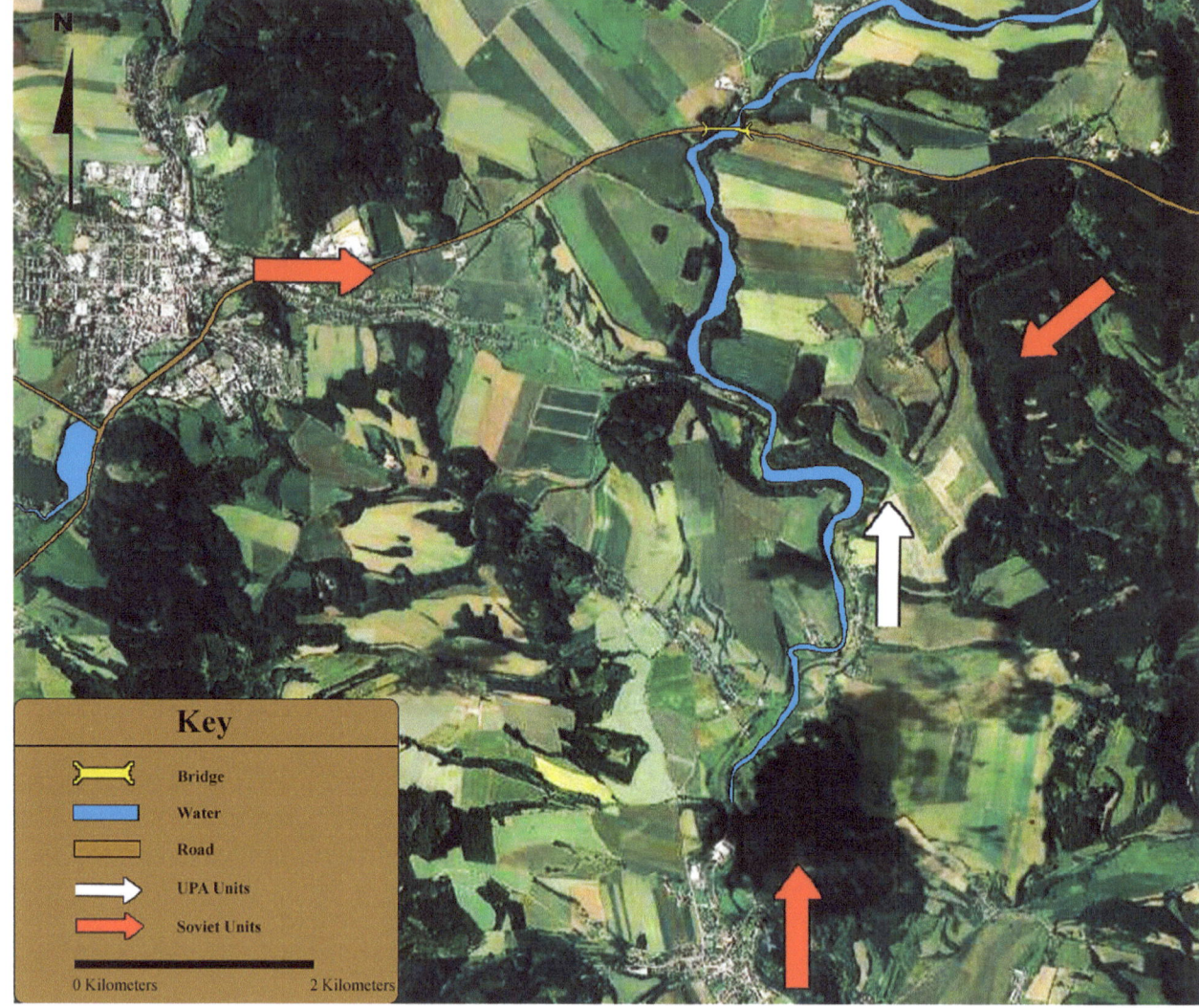

Figure 66. Scenario 9 Map.

## Game Notes

The raids of the UPA provide a variety of wargaming opportunities, from large set piece battles to small break outs. As the UPA sought to avoid direct military confrontations with the Czech forces, any tabletop recreation of these raids has one thing in common – UPA units trying to avoid encirclements. If the UPA is unable to avoid an encirclement, then the focus is escape via a breakout.

Photographs of UPA fighters in Germany show an overwhelming preponderance of SMGs. A few rifles and a light machinegun may be included in the inventory, but these would be rare. Small arms carried during the raid were almost exclusively Soviet, as ammunition for these weapons could be more readily replenished from captured enemy stocks.

In the immediate post war years, the Czech troops were equipped by the Soviet Union. For game purposes, Soviet troops can be used in place of specific Czech forces.

In this engagement, three squads of UPA fighters have been cornered by four squads of Czech troops. The UPA player has the option of breaking each squad into two fireteams at the start of the game. If using Warlord's *Bolt Action* rules, the splitting of a squad would provide an addition action die; if all three squads split into six fire teams, then the UPA player would receive six action dice.

Figure 67. 28mm UPA insurgents, author's collection.

## Order of Battle

### Czech
Commander + 2 men, Regular
Soviet Advisor + 1 man, Veteran

Squad #1, Veteran
10 men, NCO and 3 men armed with SMGs, 1 LMG, 5 rifles

Squad #2, Regular
12 men, NCO and 3 men armed with SMGs, 1 LMG, 7 rifles

Squad #3, Regular

# Bandera's Boys

10 men, NCO and 3 men armed with SMGs, 1 LMG, 5 rifles

Squad #4, Inexperienced Shirkers
12 men, NCO and 3 men armed with SMGs, 1 LMG, 7 rifles

MMG Team, Regular

## **UPA**
Commander + 1 man, both are Veterans

Squad #1, Veteran Guerrillas
12 men, NCO and 9 men armed with SMGs, 1 LMG

Squad #2, Veteran Guerrillas
10 men, NCO and 9 men armed with SMGs

Squad #3, Veteran Guerrillas
10 men, NCO and 7 men armed with SMGs, 1 LMG

Figure 68. Personal items and field gear of UPA insurgent *Zalizniak*, who in 1948 crossed into Saltsburg, Austria, from *Armia Bezsmertnykh*.

<u>Victory Conditions</u>
    The UPA player needs to get his troops off the table. For every man that reaches the table edge, the UPA player gains one point. The Czech player needs to destroy the UPA. For every UPA casualty, the Czech player gains one point and for every captured UPA player, the Czech player gains two points. The player with the most points at the end of the game wins.
    The game plays until the UPA player is destroyed or escapes.

# Scenario #10, Escape from the Bunker, March 1948

*Overview*

Winter is often a very hard time for any military force that is waging an unconventional war. Food supplies are not readily available and in areas of snow, leaving tracks can be fatal. In the first few years of the UPA, insurgents took shelter in villages not occupied by the enemy. Barns, outbuildings and sheds provided shelter for squads and companies. Armed guards would serve as sentries and maintain patrols on the village outskirts. When, due to increased enemy patrols, it was no longer possible to stay in villages, the insurgents passed the winter in the forest where they built living quarters and huts to house food, arms, and medicine.

After 1946, however, due to increased pressures from Soviet security forces, the UPA was forced to adapt to survive the winter months. For increased survivability, the units were broken down into the smaller groups. Those groups who managed to survive the winter would reform in springtime.

One option to survive the winter was continue maneuvering, which required the troops to march all day and then set up makeshift shelters of tree branches and brush. While the makeshift shelters provided the insurgents with adequate shelter from wind and snow, there were unable to provide sufficient warmth in the subzero climate. As a result, the poorly fed troops suffered from exhaustion and stress. The second option was for UPA units to construct and hide in underground bunkers.

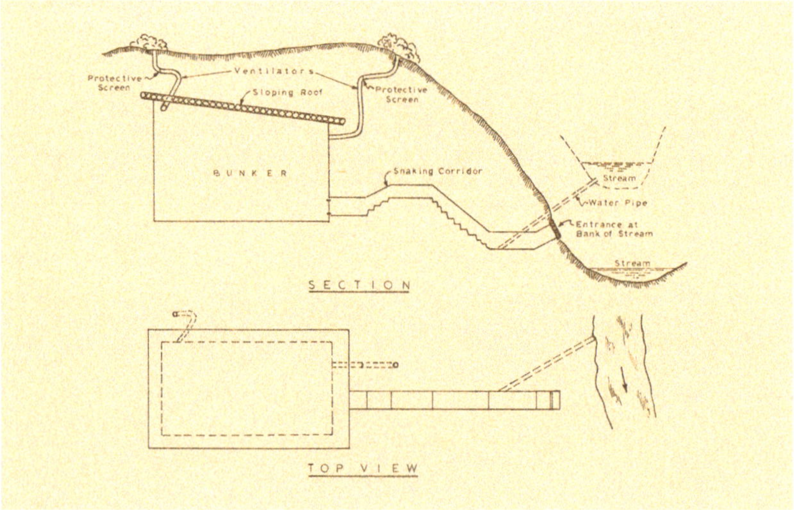

Figure 69. Plan of UPA Bunker, from *UPA Warfare in Ukraine*.

*Introduction*

After 1945, the forest of Volhynia and the Carpathian Mountains were covered with large numbers of underground bunkers. Every fall, UPA soldiers would build new structures. Building materials, such as brick, stone, and timber needed to be acquired and brought to the site. The size of each bunker depended on its purpose: some bunkers were built to house only a few men, while others could house a few dozen. Bunkers were also constructed for a particular function, including a variety of medical facilities: first aid stations, surgery centers, long-term medical recovery centers, or storage.

Whatever their function, all bunkers needed to be well hidden and camouflaged. Bunkers usually had two entrances, so that in the event an enemy attacked one entrance, those inside could flee out the second exit. Bunkers also used ventilators to provide air to those inside. Water was often brought into a bunker by a variety of means, including pipes and underwater streams. Bunkers were always built in a zig-zag fashion, so that the enemy would be unable throw hand grenades onto the insurgents below. The entrances to the bunkers were often mined and the top of the bunker usually contained a point from which those below could observe their surroundings.

Throughout the late 1940s, the Soviets and their agents focused on discovering the UPA members hiding in their bunkers. Dogs, local collaborators, and gas were often used to find a bunker. Once a bunker was found, few options existed for those trapped inside. Most often the insurgents tried to resist, but escape was difficult, especially if the insurgents had been hiding in the bunker for any length of time. Often those inside killed themselves rather than be captured and tortured by the enemy. Once a bunker was captured, the Soviets went to great lengths to document their construction.

Life in a bunker underground was difficult. Days were spent sleeping and nights were spent between KP and guard duty. Monotony was punctured by moments of sheer terror, as enemy patrols periodically passed overhead, searching for their elusive quarry. By March, the snows had usually melted and the UPA personnel emerged from their bunkers.

Figure 70. Living history event, Veretskyj Pass, Carpathian Mountains, 2018, courtesy of Jurko Datsko.

Deadly pale and emaciated by the time spent underground, accounts by UPA fighters state that the fresh air was hard to breath, and it often took a week or two for the men to recover their strength.

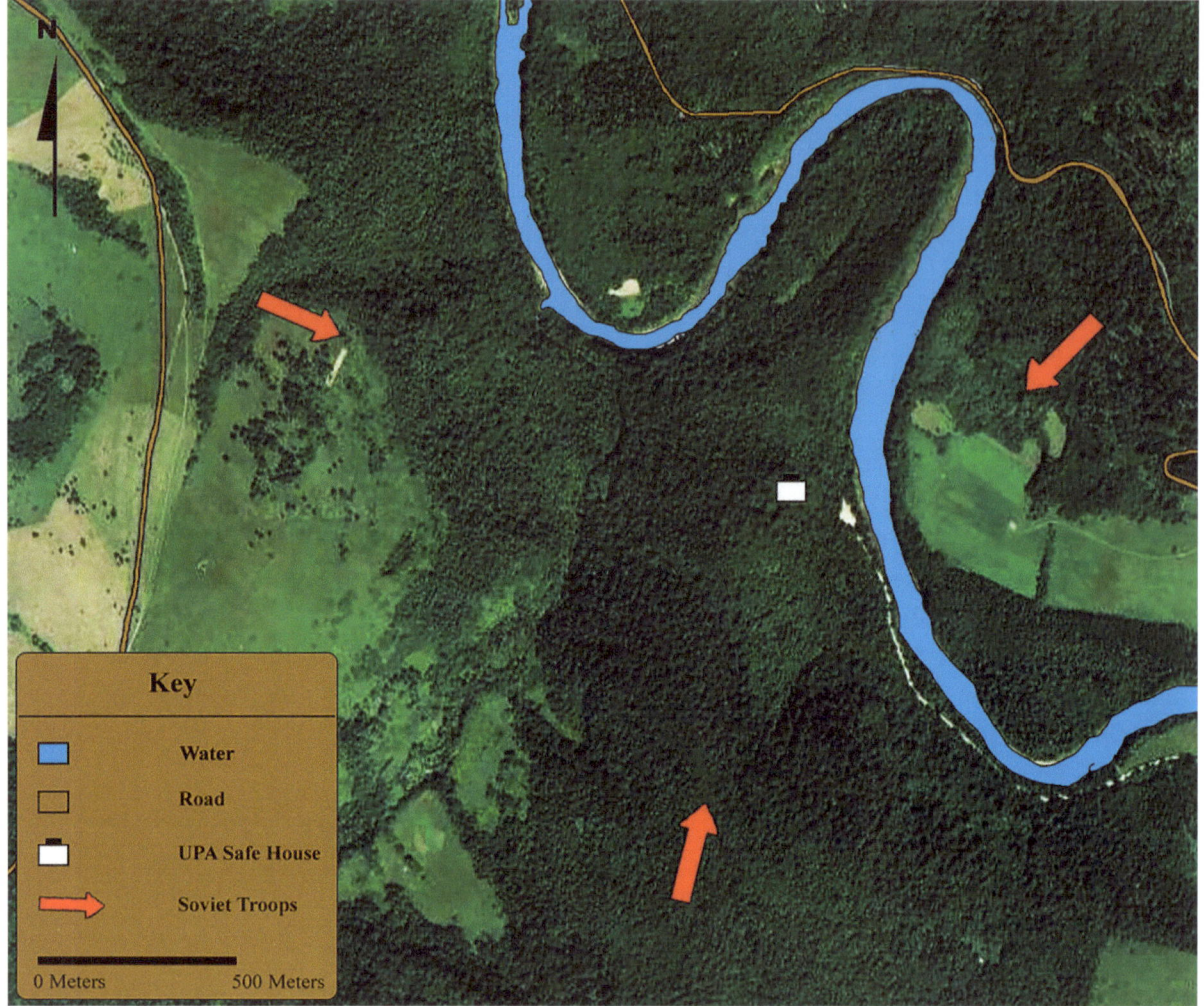

Figure 71. Scenario 10 Map.

## *The Engagement*

During the winter of 1947/48, UPA commander Stepan Khrin survived the cold months in a bunker with a handful of men. As the snow began to melt, the bunker was inundated with water. Khrin's men went outside and formed a human bucket chain to bail out the water. This proved to be fruitless and the men were forced to use picks and shovels to dig a drainage trench toward the nearby stream. Watching from on top of the bunker, the men took all of their equipment out to dry in the sun, while a fire was used to dry the remaining vegetables that were previously kept in the bunker's root cellar. The weather was still bitterly cold and many of the men had a hard time being outside of any length of time.

The following day, at about 10:00 in the morning, the UPA insurgent on watch noticed a group of Soviet troops quickly moving toward the bunker from three directions. The guard used an iron bar to strike three sharp blows – the signal that the enemy was approaching. Khrin and his men emerged from the bunker's two exits and began to disburse. Some of the men did not have the strength to run in the snow. Khrin recounted the sounds of shooting back at the bunker. After running for a kilometer and a half, the surviving insurgents were unable to move further and took cover. Those fighters who had been out on patrols the previous week felt a bit better, but even they were no condition to attack the enemy. After a short rest, the insurgents climbed a small hill to spy upon the enemy below. Unable to track the insurgents, the Soviet troops returned to the now deserted bunker.

Commander Stepan Khrin continued to lead his men and in October, he once again descended below ground to spend another winter in another bunker. To keep his mind alert, Khrin spent his time re-writing his memoirs, which were then smuggled to "The West". His memoirs were published in 1950.

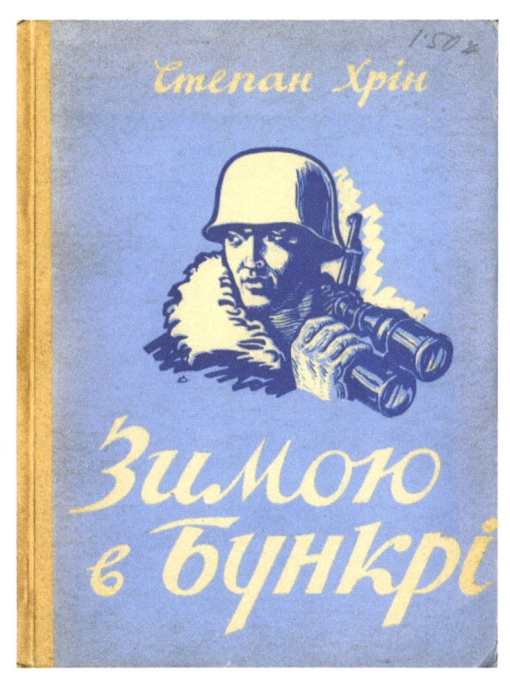

Figure 72. Cover of Khrin's memoirs, *Zymoiu v Bunkri*, author's collection.

## Game Notes

In our recreation of the escape from the bunker, the UPA insurgent on guard duty has just noticed the quickly approaching NKVD/MVD troops and has sounded the alarm. The Soviet figures are placed on three table edges and the UPA player must try to extract his men as quickly as possible toward the fourth table edge. The bunker needs to be represented by two exit points, fourteen inches apart. At the start of the game, when an UPA command die is drawn, as per Warlord's *Bolt Action* rules, a D6 is rolled for that unit. The resulting number indicates the number of men that were able to get out in that turn. If the die roll was an even number, then the figures all appear at one of the two exits and if the die roll was an odd number, then the figures all appear at the other exit. The UPA player can then move the unit or chose to stay and wait for more men to exit.

If an NKVD/MVD unit is destroyed, an additional NKVD/MVD squad of 6 men (NCO and 5 men with SMGs) appears at the table edge furthest away from any UPA unit the start of the following turn.

Bandera's Boys
## **Order of Battle**

**Soviet**

    Commander with 2 additional men, Veteran, all armed with SMGs

NKVD/MVD Squad #1, Veteran
    6 men, NCO and 5 men with SMGs

NKVD/MVD Squad #2, Regular
    6 men, NCO and 5 men with SMGs

NKVD/MVD Squad #3, Inexperienced
    6 men, NCO and 3 men with SMGs, 2 men with rifles

Flamethrower Team, Regular

Figure 73. 28mm UPA insurgents, author's collection.

**UPA**

Commander Khrin with one additional man, both armed with SMGs, Veteran
Medic, Veteran
Squad #1, Veteran
    6 men, NCO and 5 men with SMGs

Squad #2, Veteran
    6 men, NCO and 5 men with SMGs

MMG Team, Veteran

Figure 74. Melodia the Beskid sub-district medical officer (middle) with his nurse, security and wounded, Fall 1946, from *UPA Warfare in Ukraine*.

## *Victory Conditions*

The UPA fighters need to escape from their pursuers. The UPA player gains two victory points for each figure moved off the table and one point for each figure that is out of the Soviet weapon range at the end of the game.

The game plays for 6 turns.

Figure 75. 28mm Soviet NKVD/MVD figures, painted by Kostiantyn Pinaev, courtesy of Trizub Miniatures.

# Scenario #11, Collect the Drop, September 1949

*Overview*

At the start of the Cold War, US intelligence was unprepared to deal with the Soviet system. Despite conducting a number of successful operations during the Second World War, Washington was significantly outclassed by their Soviet counterparts. American cooperation with British intelligence proved to be detrimental in the long run, as MI6 was compromised by Soviet double agents. Kim Philby, arguably the most infamous of the "Cambridge Five", provided much information to the Soviets when he served as the chief British liaison officer with the American intelligence agencies.

*Introduction*

In 1945, President Harry S. Truman abolished the Office of Strategic Services (OSS), as well as other agencies created during the war. Recognizing the need for a national intelligence service in the rapidly changing post-war environment, the following year administration officials created the Central Intelligence Group which became the Central Intelligence Agency in 1947.

Among the critical tasks of the CIA was to collect information on the Soviet Union. As UPA fighters began appearing in Western Europe, the CIA began to reach out to the political leadership of the OUN. Stepan Bandera, now in Western Germany, rebuffed attempts at co-operation with the US, but a dialogue was established with another prominent OUN leader, Mykola Lebed.

Figure 76. Kim Philby, 1955.

The CIA (and its predecessors) conducted a number of operations "to exploit anti-Soviet resistance groups (based in) Western Europe for intelligence purposes"[12]. A number of such operations existed, including Trident, Cartel (pre-1949), Androgen, AE Carthage, and Aerodynamic. Both the OUN leadership and the US agents benefited from this cooperation. At times, the OUN leadership was able to unmask Soviet agents and arrange for their liquidation. The US acquired intelligence on Soviet forces in Western Ukraine and the OUN was able to drop a limited amount of supplies to their forces still operating in the field. Among the unclassified documents uncovered by the author in the US National Archives were references to the aid-drop of hand grenades and M-2 paratroop carbines. Other documents refer to the training of US operatives and the

[12]Memorandum for Chief, Plans. Project Aerodynamic (Renewal). Online at *https://cryptome.org/2016/01/cia-ua-aerodynamic.pdf*. Accessed 1 August 2022.

equipment used when they were inserted into the USSR. Unfortunately, because of the actions of Soviet agents like Kim Philby, many of these operatives were caught when they landed.

Figure 77. Airdrop.

Not all the airdrops into Ukraine were intercepted. The Soviets often photographed captured UPA arms and a US M2 paratrooper carbine appears alongside of other UPA weapons collected from Bujnyj's Bunker. Bujnyj's Bunker was seized in 1951 in Volyn'. Another photograph, this one from July 1950 shows Vasyl' Kyk, the *de facto* commander of the UPA after the death of Roman Shukhevych, carrying a US M2 paratrooper carbine. Other photographs of captured weapons show US paratrooper carbines with either fifteen or thirty round magazines. All known photographs of paratrooper carbines show them with the late war integral bayonet lug and adjustable rear sights. Though it is possible that some of these carbines were brought by individual couriers moving back into Ukraine, the documents in the US National Archives are clear that a number of US provided weapons, including M2 carbines, ammunition, and hand grenades were dropped to the Ukrainians. A Soviet photograph of the recently killed OUN operative "Jurko" that had parachuted near the city of Bolekhiv in 1952 shows him wearing what appears to be US made Mk2 A1 "Pineapple" fragmentation grenade[13].

---

[13]Shchehliuk, Myron. 434.

Figure 78. Scenario 11 Map.

## *The Engagement*

A single twin-engine aircraft has successfully crossed over into Soviet territory on the way to drop an operative and a cargo of supplies. The Soviets, aware of the aircraft and its cargo, have sent ground troops to deal with the capitalist incursion. The OUN operatives, alert to the overwhelming Soviet ground presence at the drop zone, were powerless to stop the drop and can only watch from the relative safety of a nearby mountaintop.

At this moment, fate intervened as one of the aircraft's engines began to sputter. Rather than risk continuing to the predetermined drop zone, the aircrew chief made the call to drop the man and the packages earlier. The door opened and the man, along with his cargo, began to fall to the earth. Before the first parachutes opened, the airplane was turning around and heading for home.

At this point, three small squads of former UPA fighters race to get to the man before the Soviet authorities can reorient themselves and begin to track the enemies of the USSR.

Figure 79. 28mm UPA couriers, author's collection.

## Order of Battle

**Soviet**
see Games Notes below

**UPA**
Commander + 1 man, both armed with SMGs

Squad #1, Veteran Guerrilla Fighters
NCO and 4 men, all armed with SMGs

Squad #2, Veteran Guerrilla Fighters
NCO and 4 men, all armed with SMGs

Squad #3, Veteran Guerrilla Fighters
NCO and 4 men, all armed with SMGs

Figure 80. 28mm Soviet NKVD Squad, courtesy of Warlord Games.

## Game Notes

In this historical "What If" scenario, a small handful of diehard UPA fighters continue to wage war of the Soviet system. Large scale warfare has become a thing of the past and those that remain spend most of their time waging a propaganda campaign of writing, printing and distributing leaflets. At times an OUN cell will conduct an impromptu assassination of a local party member but fear of Soviet retribution on the local civilian population remains high. Few civilians can be trusted, as some former OUN fighters now work for Moscow.

In this game, the three UPA groups do not know each other and have little trust for those who proclaim to work for Ukrainian independence. Each of the four groups must try and find the courier and take possession of the supplies. Those supplies that cannot be secured must be destroyed.

The Soviets are scrambling at the start of the scenario. What was to be an easy intercept, thanks to the information passed on by their British agent, has become a clusterf**k. While many heard the sound of the aircraft's engines, a watchful NKVD sentry had spotted the parachutes deploying a few kilometers to the south. As a result, all Soviets units were proceeding with all possible speed to the drop zone.

At the start of the game, the courier and the three supply boxes are scattered in the center of the game table. The courier has suffered a sprained ankle and can only move a half speed. Not sure where to go, he will wait until the first UPA unit reaches him. Once contact is made, he will follow the commands of the Ukrainians. If Using Warlord's *Bolt Action* rules, the courier has his own command die.

At the start of each turn, a D20 is rolled three times to see which Soviet units appear on the north side of the game table (See Soviet Appearance Table below). A particular Soviet unit can only appear once during the game. Thus, if on a following turn, the same number appears on a subsequent D20 roll, no additional troops are added to the game.

*Soviet Appearance Table (D20)*

1. Veteran Medic + 1 assistant, both armed with pistols
2. nothing
3. nothing
4. GAZ Truck with Inexperienced NKVD Squad of 6 men, armed with 3 SMGs and 3 rifles
5. nothing
6. nothing
7. Regular NKVD Squad of 6 men, armed with 3 SMGs and 3 rifles
8. nothing
9. nothing
10. Regular NKVD Captain + 2 men, all armed with SMGs
11. nothing
12. nothing
13. Veteran NKVD Squad of 6 men, armed with 3 SMGs and 3 rifles
14. nothing
15. nothing
16. Lend-Lease jeep with Inexperienced NKVD Light Mortar Team
17. nothing
18. Veteran NKVD Squad of 6 men, all armed with SMGs
19. nothing
20. Inexperienced NKVD Lieutenant + 2 men, 1 pistol and two rifles

## Victory Conditions

The side which ends up with the courier and gets him off their side of the table wins the game. If it appears that the enemy has taken control of the parachutist, it is perfectly appropriate in game turns to try and kill him. If the courier is killed, the side which controls the majority of the dropped supplies wins the game.

## Scenario #12, Prison Uprising, May 1954

### Overview

At its inception, the Soviet system used concentration camps. The corrective labor camps, better known as Gulags, were initially set up in every province but by 1920, the more dangerous political prisoners were sent to the Solovets Islands. In the late 1920s this network expanded and prison labor was used to build the White Sea Canal, hydroelectric station, and railroads. In the 1930s, over two million intellectuals and farmers from Ukraine were sent to these facilitates and were used as slave labor. To control these political prisoners and class enemies, gangs of violent criminals were also incarcerated in the same facilities. These gangs terrorized and exploited the intellectuals, occurrences of brutality, cruelty, sadism, and rape were common. By 1941, fourteen percent of the economic output of the USSR was produced by prisoners.

In 1945, as Soviet troops moved toward Berlin, new groups of prisoners began to arrive in these camps. In addition to the German POWs, insurgents from Western Ukraine, Western Belarus, and the Baltic nations were also sent to the GULAG. Under the control of the Ministry of State Security (MGB), the new political prisoners were a different creature from the intellectuals and nationalist that came before. The violent gangs that used to prey upon the prisoners were now put in place by the former UPA insurgents and the Lithuanian "Forest Brothers". Though of different ethnic background, the incarcerated combatants from Ukraine, Lithuania, Latvia, Estonia, Belorus, Georgia, and Chechnia all had fought against the Soviets and would cooperate against a common enemy.

Figure 81. Kingir Camp, courtesy of Jurko Datsko.

# Adrian Mandzy

In 1953, Stalin died and his right-hand man, Lavrentiy Beria, the head of the secret police, was arrested and executed. For the prisoners continued within the camps, hope sprung for a better future. At the same time, the leadership of each camp was uncertain of their future, as they all had connections of some sort with Beria and his appointed administration. The sometimes-arbitrary killing of prisoners by the camp guards, for no apparent reason, added to the tension within the camps. As the violence within the camps increased, the prisoners in many camps took action. Strikes were common, and as they were legal under the Soviet system, the guard and camp leadership were unclear how to respond. As the number and duration of strikes increased, so did the guard's frustrations.

## *Introduction*

While strikes and riots occurred previously at other camps, only the prisoners at the Kingir Camp in Kazakhstan were successful in driving out the Soviet authorities. In the weeks that followed, prisoners within the confines of the camp not only set up an elaborate system of government, with technical teams, a defensive military wing, a food department, services (such as laundry, and shoe repair) and an internal security organization that monitored any prisoner who wanted to surrender to the Soviet authorities. Though Ukrainians dominated the camp, the leadership was politically savvy enough to choose a Russian Red Army officer as the face of the opposition. Turning the words of Soviet propaganda against the authorities, the inmates hung banners stating "Long live the Soviet Constitution!" and "Down with the murdering Beria-ites!".

Figure 82. Kingir Prisoners, courtesy of Jurko Datsko.

Within the confines of the camp, the prisoners sought to create a better world for themselves. Food and clothing supplies, previously taken by the guards and place in storerooms, were redistributed to the inmates. One enterprising old aristocrat even managed to open a coffee shop which served substitute coffee.

Given the large number of playwrights and artists incarcerated within the prison, the former inmates began to preform plays and sing new songs celebrating their newfound freedom. The formerly imprisoned clergy also became active and conducted services.

During their forty days of autonomy, the inmates waged a type of cold war with the outside authorities. Radio broadcasts, leaflets, placards, and even kites all sought to show the rioters as not anti-Soviet, but rather as proper soviet citizens who were being denied their rights. Among the demands were that guards responsible for the deaths of inmates be held accountable, that they no longer had to wear prison numbers, that an eight-hour workday be put in place, and most importantly, that their cases be reviewed.

The Soviet authorities were not interested in negotiating with the prisoners and explored different ways of ending the current standoff. Initial attempts to sow internal discord proved unsuccessful and while a few loyal prisoners escaped to the authorities, those who spoke of surrender were placed in the camp's jail. To give the inmates a false sense of security, Soviet authorities promised over a loudspeaker that they would meet with a member of the Soviet Central Committee.

Figure 83. 54mm Soviet T-34 and troops, courtesy of Tod Kershner.

The following day the Soviets sent in the troops to crush the uprising. Led by five T-34 tanks and supported by close to two thousand troops, the attackers swiftly broke through the barb wire enclosures surrounding the camp. In the panic and chaos that followed, Soviet troops shot randomly. Though one tank had become immobilized when it tried to cross a line of latrines, the inmates, armed with a few small arms, iron bars, and improvised grenades, were unable to stop the onslaught. Hundreds of people lost their lives - either by being run over by the tanks or being machinegunned. Special squads were assigned to capture the ringleaders of the uprising alive, so that after a special trial they all could be publicly executed. Ninety minutes after the assault began, Soviet authority was reestablished in the camp.

Figure 84. 28mm rioting prisoners, author's collection.

## *The Engagement*

At the times of the uprising, the camp held a little more than 5,000 inmates. Of these, seventy percent of those incarcerated were either Ukrainian or Balts who had fought in the underground against the Soviet authorities. Women, who were kept in a separate section of the camp, made up close to 43 percent of the prison population[14]. Though separated from the men's barrack by a wall, messages were often exchanged between the two groups.

The uprising began on 16 May 1954. That evening, a group of inmates had tried to break into the central courtyard, where food was stored, and then into the

---

[14] Steven Barnes, "In a Manner Befitting Soviet Citizens": An Uprising in the Post-Stalin Gulag. *Slavic Review*, Vol. 64, No. 4 (Winter 2005), 829. Article is on pages 823-850

women's section of the camp. They were soon chased away by the guards but returned a few hours later. Using an improvised battering ram, the prisoners broke down the walls that separated the various sections of the camp. The guards responded by opening fire on the prisoners, killing a handful and wounding a few dozen more. More prisoners joined the uprising and the guards withdrew.

The following day, when the prisoners went to work outside the camp walls, the guards began to repair the walls separating the separate compounds. A few hours later, when the prisoners returned from work, they saw the recently repaired walls. Negotiations with the camp guards proved unsuccessful and after a failed cease-fire, the inmates responded to the violence by attacking the guards within the confines of the camp. With casualties on both sides, the guards withdrew behind the wire.

Figure 85. Scenario 12 Map.

## Game Notes

The game focuses on the capture of the camp by the prisoners. Historically, the camp guards withdrew when challenged by the inmates, but in our recreation, the prisoners need to "motivate" the guards a bit more before they abandon the central open area. Casualties will occur during the game, as happened in real life, but neither side has the possibility of eradicating the other group. The inmates do not have the means at their disposal (shives and clubs do not do well against SMGs) and the guards do not have the luxury of carrying our mass exterminations for two reasons. The first, is that the prisoners are required to fulfill work quotes and dead men do no digging. Secondly, the change of political leadership may result in a series of reports, followed by direct questions from Moscow bureaucrats, which in turn may lead to a firing squad.

The guards must drive the inmates out from the courtyard. They are outnumbered and though they have a two machinegun towers that overlook the camp, they cannot kill more than ten percent of the prisoners. For this reason, the fire from the tower will cause at most only one casualty per firing sequence and pins (if using Warlord's *Bolt Action* rules). When engaged in close combat, the losing group needs to retreat and should suffer only minimal casualties (one to two figures). If an armed guard will lose to a prisoner, the inmate (or inmates) may has/have possession of a firearm and can use it the following turn. The guards have no restrictions on shooting armed prisoners.

Figure 86. 28mm Soviet NKVD/MVD figures, painted by Kostiantyn Pinaev, courtesy of Trizub Miniatures.

If the rule set you are using for this scenario does not include a section on hand-to-hand combat, the following rules can be used. The attacker moves in on the defender, presuming they have enough movement to close the distance. If the unit has pin markers or is disordered, follow your rules to see if they need to pass any type of moral check to close with their target. Once in close combat, count the number of figures in your group (one point each). If one side has more miniatures in the melee than the other side, add two points for every addition figure. Each side rolls a D6 and adds the points for the figures in their group. In case of a resulting tie, both sides lose a figure and roll again. The highest total number wins the melee. The loser then retreats a full movement away from the enemy. The defeated group also loses at least one figure. If the attacker's number doubled the defender's number, then the defender loses two figures. If a group of prisoners defeat a group of guards, they are now considered to be armed with the weapons of their former oppressors.

Movement is restricted between the three areas of the camp. To move from one area into another, a player needs to spend an action to move from one compound into another.

Figure 87. 28mm armed former prisoners, author's collection.

Adrian Mandzy

# **Order of Battle**

## **Camp Guards**
Leader +2 men, Regular

Squad #1, Veteran
NCO + 9 men, 1 armed with SMG and nine with rifles

Squad #2, Regular
NCO + 9 men, 1 armed with SMG and nine with rifles

Squad #3, Regular
NCO + 9 men, 1 armed with SMG and nine with rifles

Squad #4, Inexperienced
NCO + 9 men, 1 armed with SMG and nine with rifles

Tower #1
MMG, firing at long range

Tower #2
MMG, firing at long range

## **Inmates**
UPA Squad #1, Veteran Guerrilla Fighters
NCO + 15 men, all armed with knives and clubs

UPA Squad #2, Veteran Guerrilla Fighters
NCO + 15 men, all armed with knives and clubs

Lithuanian "Forest Brothers" Squad, Veteran Guerrilla Fighters
NCO + 15 men, all armed with knives and clubs

Latvian Squad, Veteran Guerrilla Fighters
NCO + 15 men, all armed with knives and clubs

Chechen Squad, Veteran Guerrilla Fighters
NCO + 15 men, all armed with knives and clubs

Belorussion Squad, Veteran Guerrilla Fighters
NCO + 15 men, all armed with knives and clubs

# Bandera's Boys

Criminal Gang, Inexperienced, Shirkers
12 men, all armed with shivs and iron rods

### *Victory Conditions*

For the inmates to win, they will need to drive the camp guards out of the camp. Conversely, the guards must control the camp courtyard – no prisoners can be in the open space. If neither side controls the courtyard at the end of ten turns, the game is considered a draw.

# Appendix #1
*Weapons of the Ukrainian Insurgent Army*

As weapons and ammunition needed to be acquired from the enemy, the armament of the Ukrainian Insurgent Army was, in a word, varied. As ammunition was not always interchangeable, weapons acquired from one foe would often no longer be usable when another enemy appeared. After action reports filed by UPA junior officers discussed how much ammunition was used in each encounter and those fighters who were not conservative with their use were subject to punishment.

Before the start of WW2, the OUN maintained small cashes of weapons to use in their attacks on Polish and Soviet officials. As the weapons needed to be concealed before the start of an attack, pistols and cut down old military rifles were the most commonly used firearms. Hunting arms, such as shotguns, were also employed in limited numbers. Explosives, including both homemade devices and hand grenades acquired from military arsenals, played a large role in the OUN's attacks. A cache of weapons collected from the OUN by the Polish police in 1933 includes a number of modern small caliber pocket handguns and service side arms from WWI, including a Roth Steyr M1907, a Frommer Stop, and a Rast & Gasser.[15]

Figure 88. OUN cache of arms captured by the Polish Police, 1933, from *Strilets'ka Zbroia Ukrains'kukh Povstantsiu*.

During the period of the short-lived Ukrainian Carpathian Sich, OUN volunteers travelled to Khust in Czechoslovakia in an effort to halt the Hungarian occupation. Arms used during this period came from former Slovak arsenals as well as weapons brought in from Poland. In the long run, however, these proved to be inadequate for the task at hand. OUN documents and memoirs often mention the use of rifles and revolvers, but no additional information is provided as to type, country of manufacture, or any other characteristics.[16] Photographs and newsreels of this period show the Ukrainian fighters using

---

[15] Myron Shchehliuk, *Strilets'ka Zbroia Ukrains'kykh Povstantsiv/ Firearms of the Ukrainian Partisans/*. L'viv: Spolom. 2014. 103.

[16] Shchehliuk, *Strilets'ka Zbroia Ukrains'kykh Povstantsiv/ Firearms of the Ukrainian Partisans/*. 126.

the ZB26 machinegun and Mauser rifles. Scenes from the 1939 film "Silver Land", which were shot on location, show battle scenes with actors using C-96 "Broomhandle" Mauser pistols, a German MP34/1 submachinegun, and Austrian M95 rifles and carbines.

Figure 89. Austrian produced M95 carbine, author's collection.

Following the German and Soviet invasion of Poland, Ukrainians serving in the Polish army began to collect arms for the OUN. Rifles, such as the Polish version of the German Mauser, entered the Ukrainian arms caches. Pistols, like the Polish VIZ-35 "Radom", were highly prized and sought after. Battlefields from the short conflict were also searched and were appropriate, arms and more importantly, ammunition, were collected and stored for the upcoming struggle. Among the identifiable weapons collected from OUN members in 1940/41 during "Operation 59" were a 1903 Browning, a Nagant revolver, a Sauer 38H, and a German M34 grenade.[17]

In June 1941, the Nazis embarked on their great crusade against communism. Within weeks of their attack, the Soviets had lost millions of troops and tons of equipment. Outside of L'viv, an abandoned Soviet T-35 tank was decorated with a large trident and an epithet praising Stepan Bandera.[18] Ukrainians who sought to create independent local government organizations made use of abandoned Soviet gear, including weapons. The new Soviet semi-automatic SVT rifles were highly

---

[17] Shchehliuk, *Strilets'ka Zbroia Ukrains'kykh Povstantsiv/ Firearms of the Ukrainian Partisans/*. 114.
[18] In 1998, the Ukrainian company ICM released a 1:35 scale model of the T-35. Among the decals provided in the kit were marking for the "Ukrainian" version of this tank.

sought after, as were submachineguns like the PPD, but any weapon, including older bolt action rifles, were acquired and put away in anticipation of the upcoming struggle.

Figure 90. UPA insurgents armed with SVTs, from Iavorivs'kyj Fotoarkhiv UPA.

In 1943, as the OUN began to expand into the UPA, the need for weapons increased exponentially. The previously prized weapons were ones that could easily be hidden – pistols, cut down rifles, submachineguns, and grenades. As the battles fought by the OUN in defense of the Ukrainian Carpathian Sich, as well as the military campaigns fought throughout Ukraine in 1941 demonstrated, new types of weapons were needed. The UPA squad was designated to have one or two members with submachineguns and a few men with automatic weapons. Integral to the squad's operation was a machinegun, which was to be supported by a handful of additional men armed with bolt action rifles. Additional weapons, such as heavy machineguns and mortars, could be used to either support UPA platoons and companies as needed for a particular mission, or they were grouped together to support larger set piece battalion size actions.

## Submachineguns

Submachineguns were useful weapons for insurgents. Relatively light, somewhat dependable and relatively accurate at close ranges, these weapons could put out a tremendous amount of firepower in a very short period of time. Photographs show insurgents using a variety of both 9mm and 7.62mm SMGs, including German MP28s, MP38s, and MP40s, Soviet PPDs, PPSh-41s, and PPS-43s, and in smaller numbers, the Italian Beretta Model 38, the Czech ZK-383, and the Romanian Orita 1941. A wartime photograph of Myroslav Symchych shows him armed with a PPS-43 Soviet submachine gun. When asked about his weapon, Symchych stated that he had modified his weapon by replacing the "piece of junk" barrel with a new one he had taken from a recently captured Soviet rifle. According to the former insurgent, he had shortened the barrel by cutting it in half.[19]

Figure 91. Myroslav Symchych, on the left, with his "improved" PPS-43, courtesy of Jurko Datsko.

One photograph shows a UPA insurgent with an American Thompson M1A1, but as ammunition for this weapon was uncommon in Ukraine, and the weapon was significantly heavier than other SMGs, it is unclear how long was the weapon remained in use.[20] As submachineguns were difficult to acquire during the transition of the OUN into the UPA, photographs show entire squads of insurgents equipped with SVT rifles, which may have functioned as assault squads.

**Rifles**

Submachineguns, however, did have their drawbacks. Not only did they require the insurgents to carry around a large quantity of ammunition, which was never easy to replenish, but they were inadequate in dealing with an enemy at longer ranges. To address these issues, UPA units began to increase the amount of men armed with rifles in each squad.[21] Though bolt action rifles were becoming increasingly obsolete throughout the 1940s, almost all of the major world powers relied on them. The Soviets used a variant of the Mosin-Nagant 3-line rifle. Originally developed at the end of the 19th century, the Mosin rifle remained in production by the Soviets until the end of WWII, while the M44 carbine version of the Mosin was produced until 1948. Similarly, the Germans relied on the Mauser 98k and the Hungarians continued to field a carbine version of the Steyr M95 rifle. All of these weapons were used by the UPA insurgents.

---

[19] Myroslav Symchych, Interview with Adrian Mandzy, on the site of the Kosmach Ambush, 4 July 2011.
[20] Shchehliuk, *Strilets'ka Zbroia Ukrains'kykh Povstantsiv/ Firearms of the Ukrainian Partisans/*. 274.
[21] Volodymyr V'iatrovych et al., *Ukrains'ka Povstans'ka Apmia: Istoria neskorenykh/ Ukrainian Partisan Army: History of the Undefeated/*. L'viv: Tsentr doslidzhennia vyzvol'noho rykhy. 2011, 184.

Figure 92. Spent Hungarian cartridges marking an insurgent's firing position were recovered during the author's Battlefield Survey of Kosmach, 2012.

UPA field and combat manuals discuss the use of snipers.[22] Though evidence for UPA snipers remains limited, it is believed that they were employed as a matter of course during field operations. The weapons used by the snipers, some of which were trained before the war, would have been either civilian hunting firearms or captured military weapons. The manual discusses the role of a sniper and his obligations, which includes "confidently removing a target with one shot, always maintain the weapon and optics in top condition, and use the terrain as a mask".[23] The use of Soviet sniper ammunition by the UPA was documented during the course of the 2012 archaeological survey of the Kosmach battlefield.[24]

---

[22] Four separate hard-covered volumes appeared under the title *Infantry Combat Manual (Boyovyi pravylnyk pikhoty)*. Published in 1944, the UPA manuals are a translation of a 1943 Red Army Manual published in Moscow under the same name. These manuals were reprinted in P. Sokhan, Y. Shtendera, and V. Ivanenko (eds.), *Publications of the UPA Supreme Command, Litopys UPA, New Series, Vol. 1.*, Kyiv: National Academy of Sciences of Ukraine, 1995.

[23] P. Sokhan et al., *Publications of the UPA Supreme Command, Litopys UPA, New Series, Vol. 1.*, Kyiv: National Academy of Sciences of Ukraine, 1995: 326.

[24] Adrian Mandzy, Ambush in the Mountains: A Multi-Disciplinary Study of a Successful Field Operation by a Company of the Ukrainian Partisan Army against a Soviet NKVD Battalion in January 1945, *Partisans, Guerillas, and Irregulars: The Archaeology of Asymmetric Warfare*. Tuscaloosa: University of Alabama Press, 2019.

Figure 93. UPA officers and men from the Black Forest group, May 1947. Note sniper rifle with PU scope on the lower righthand side of the photograph, from *Strilets'ka Zbroia Ukrains'kukh Povstantsiv*.

As WWII progressed, weapons design moved to provide a semi-automatic version of a battle rifle. The pre-war Soviet SVT series of weapons, produced in significant numbers before 1941, was too complex for most Soviet recruits, and required too much time and resources to produce during wartime. German troops often used captured examples of the SVTs and began to build their own versions – the Gewehr 41 and later the slightly less complex Gewehr 43. Neither were built in large numbers and were quickly superseded by the Sturmgewehr 44. This legendary weapon, often called the world's first assault rifle, used a specific cartridge, the 7.92x33mm Kurz. Though introduced too late in the war to impact its final outcome, unspecified quantities of these Sturmgewehr 44s found their way into the hands of the UPA insurgents. Unfortunately for the UPA fighters, 7.92x33mm Kurz ammunition remained rare and without access to additional supplies, these weapons did not serve for an extended length of time.

Anti-tank rifles, developed by most armies before 1939, were adopted by the Soviets following the 1939 Invasion of Finland. Though the wartime development of armor relegated the anti-tank rifle to relative obsolescence in many armies, the Soviets continued to employ and use the weapons in new ways. Unable to penetrate the front armor of most frontline combat vehicles, the relatively large caliber 14.5x114mm Soviet rounds were more than adequate to take out a soft skinned or a lightly armored vehicle. The two major Soviet anti-tank rifles, the bolt action Degtaryev PTRD and the semi-automatic Simonov PTRS, along with the German PzB39 and the Swedish produced Solothurn S18-100, saw use by the UPA on special operations. These anti-tank rifles could be deployed in groups or individually depending on the nature of the upcoming action.

Figure 94. UPA Insurgent with Sturmgewehr 44, from *Strilets'ka Zbroia Ukrains'kukh Povstantsiu*

Figure 95. Insurgents with Solothurn MG 30s, from *Ukrains'ka Povstans'ka Armia*.

## Machineguns

During the previous world war, machineguns became an integral component of the infantry. Theoretical debates on the future of warfare influenced weapon designs, which in turn dictated how field armies operated. Likewise, the UPA was influenced by how their enemy's forces were organized and operated. By the 1940s, in many armed forces, including the UPA, squads were centered around machineguns.

Light machineguns were well suited for the type of combat operations conducted by the UPA. Initially developed in the later phases of WWI, these weapons could be carried and depending on the model of the weapon, could be fired by an individual soldier on the move. Early models, like the German model MG 08/15 and British Lewis continued to be used throughout the 1940s and saw limited use by the UPA. Interwar models of light machineguns, such as the Solothurn M30, the Rheinmetall MG13, and the ZB26, which used box magazines, were often fielded by the UPA. Another inter-war model, the Soviet Degtyaryov DP-27/DP-28 light machinegun, which used a round disk magazine, well suited for use by the UPA. The weapon was relatively commonly encountered, either in its original form or as the DP-29, which was modified for use in tank. As the Degtyaryov used the standard 7.62x54R Soviet infantry cartridge, ammunition for the weapon was plentiful. The Degtyaryov's relatively low rate of fire reduced ammunition use, which was also an asset for the UPA.

Figure 96. Two UPA commanders, one armed with a Soviet submachinegun and the other with a German machine-pistol, from *UPA Warfare in Ukraine*.

The German MG34, along with its later variant the MG42, was a universal machinegun. With its high rate of fire, the portable weapon sought to combine the function of a light and a medium machinegun. Its high rate of fire of 900 rounds per minute was considered to be high enough for use in an anti-aircraft role, while its accuracy allowed for experienced solders to use the weapon in a sniping role. The weapon used the standard German cartridge and when maintained, it functioned very well on the battlefield. UPA fighters often used the MG34 and MG42, despite the weapons high rate of fire.

Medium machineguns, like the Soviet Maxim models M1910 and M1940, and its SG-43 replacement, along with the Austrian *Schwarzlose* M1907/12, the German MG08, and the Czechoslovakian ZB53/Vz37 are documented as having been used by the UPA. One photograph shows a group of UPA insurgents training on a mounted Browning M1917A1 machinegun, but its use in combat remains unclear.[25]

## Explosives and Tactical Support Weapons

In the early years of the OUN, explosives were commonly used in the strug-

---

[25] Shchehliuk, *Strilets'ka Zbroia Ukrains'kykh Povstantsiv/ Firearms of the Ukrainian Partisans/*. 358.

gle against the Polish state and representatives of the Soviet Union. Like other contemporary organizations which used terrorism to achieve their political goals, the OUN made use of home-made devices. In the conflicts that followed, grenades were commonly used by various sides. Highly sought after and cached by Ukrainians over the course of the previous decade, by 1944 a large number of grenades was available for use by the UPA. These include various types and models of hand grenades, with German and Soviet hands grenades being the most commonly used by the UPA.

Figure 97. UPA insurgents with hand grenades, from *Armia Bezsmertnykh*.

Though a hand grenade could be very lethal, its relatively short range of deployment of under 40 meters, was a limitation.[26] In the inter-war period, armed forces sought to create a longer-range delivery system that could send a grenade type explosive further. Various types of rifle grenades were developed and used by both German and Soviet forces, though they do not appear to have been commonly used by the UPA. Other anti-tank weapons that functioned on the same principle, such as the American Bazookas and the German series of single shot Panzerfaust, do not appear to have been used in any consistency by the UPA insurgents.

## Mortars and Rockets

Mortars were commonly used by the UPA during this period. 50mm Soviet mortars were common and widely available. The portability of the weapon and its comparatively light weight ammunition made it a constant companion within UPA units.[27] German 80mm and Soviet 82mm mortars were both used by the UPA, usually at the company level, but as the Soviet 82mm mortar could fire German shells designed for the German 80mm mortars, the Soviet ones were preferred. The weight of the 82mm mortar, as well as its ammunition, limited its use by the

---
[26] Shchehliuk, *Strilets'ka Zbroia / Firearms/*. 422.
[27] Volodymyr V'iatrovych et al., *Ukrains'ka Povstans'ka Apmia: Istoria neskorenykh/ Ukrainian Partisan Army: History of the Undefeated/*. L'viv: Tsentr doslidzhennia vyzvol'noho rykhy. 2011, 199.

UPA.

Rockets, such as the Soviet *Katyushas* and the German *Nebelwerfers*, sought to saturate a target area. Though not nearly as accurate as conventional artillery, the rockets delivered a number of explosives to a target area. The distinctive hollowing noise produced by the rockets when fired was unmistakable and it had a habit of terrifying those on the receiving end of a bombardment. During the course of UPA operations, both German and Soviet rockets were periodically captured and used against their former enemies. Unlike the Soviets, who mounted the rockets on trucks, or the Germans, who often fired them in groups of six from a dedicated 6-tube launcher, the UPA often fired captured rockets individually. Used in a number of UPA operations, the somewhat accurate rockets gave the insurgents a psychological, as well as a tactical advantage, over their enemies.

Figure 98. OUN leaders, UPA insurgents, and new recruits, 31 August 1944, from *Army of Immortals*. Note German (PzB-38 or 39?) anti-tank rifle in the front.

## Artillery

In 1943 and 1944, various UPA battalions fielded artillery companies. The captured guns were of various calibers and had despite the best efforts of the UPA logistics, had limited ammunition. Though used in various battles against both the Germans and the Soviets, the lack of dedicated artillery transport and the ability to move quickly hindered their wide-scale adoption. Among the guns known to have been used by the UPA were 76.2mm German guns, Soviet 45mm anti-tank cannons and 122mm howitzers. Photographic evidence shows teams of two horse pulling what appear to be limbers and Soviet 45mm anti-tank guns through the snow.

Figure 99. Insurgents of UPA-North with limbered artillery on the move, winter 1943-4, from *UPA: They Fought Hitler and Stalin.*

## Armored Cars, Tanks, Tank Destroyers, and Aircraft

Trophy weapons were such as armored cars, tanks, tank destroyers, and aircraft were periodically captured by the UPA in various battles. Though at times such weapons could be made operational, they were ill-suited for the type of war being conducted by the UPA. Vehicles continuously need upkeep and parts, which could not always be acquired. Moreover, the individual armored vehicles captured by the UPA were more of a hindrance than an asset in the unconventional war being waged. A group of armed men can attack a target and then disburse into a marsh, wood, or mountain range. You cannot easily do that a Panther or a T-34 tank.

UPA memoirs periodically talk about using captured vehicles, usually for a one-off encounter. For propaganda purposes, captured tanks could be decorated with Ukrainian markings, but such high-profile items would become a prize that an enemy would focus upon. As an immobile pill-box, a tank could provide a temporary strongpoint, but as the Germans found out late in the war, such improvised defenses were a death trap for those inside.

Though it has been documented that the UPA had captured multiple aircraft during their operations, including a Gotha Go 145 and Messerschmitt Bf 109 G-6, only the repurposed German fighter is known to have conducted an operation for the UPA. The ME 109 was captured by a young insurgent and after conducting repairs, was made operational.[28] According to the memoirs of UPA Lieutenant Colonel (*pidpolkovnyk*) Khmel, the German wing insignia was repainted with Ukrainian tridents. In June of 1944 the plane made a successful flight over the town of Dolyna and later landed without incident. The fate of the pilot and the aircraft remain unknown.

Figure 100. Captured ME 109 in Ukrainian markings, courtesy of Jurko Datsko.

**Late UPA Weapons**

By 1947, the days of large units were a thing of the past. The UPA units that continued to operate usually deployed in small groups of multiple squads. As the last reserves of ammunition acquired from the Nazi's were being used up, the UPA was relying almost exclusively on Soviet arms. For those UPA fighters ordered to march to the west, the old long-range bolt-action rifles were replaced by smaller, concealable SMGs. PPSh-41s, with their long wooden stocks, were commonly shortened by creating a pistol grip. A photograph from 1950 shows a small group of UPA couriers armed with British MkII Sten SMGs.[29] As a result of the OUN's growing contact with US intelligence, limited photographic evidence shows that at least a few OUN members were armed with M2 paratrooper carbines and US styled Mk2 fragmentation "pineapple" anti-personnel hand grenades. By the early 1950s, like the OUN operatives from the previous decade, the remaining fight-

[28] Volodymyr V'iatrovych et al., *Ukrains'ka Povstans'ka Apmia: Istoria neskorenykh/ Ukrainian Partisan Army: History of the Undefeated/*. L'viv: Tsentr doslidzhennia vyzvol'noho rykhy. 2011, 202.
[29] Shchehliuk, *Strilets'ka Zbroia Ukrains'kykh Povstantsiv/ Firearms of the Ukrainian Partisans/*. 284.

ers relied on arms that could easily be concealed and were most effective at close range.

Figure 101. Weapons and radios seized by Soviet authorities from a bunker, 1951, from *Strilets'ka Zbroia Ukrains'kukh Povstantsiv*. Note US M2 paratrooper carbine in the middle.

# Appendix #2
*Uniforms of the Ukrainian Insurgent Army*

Like the weapons, the uniforms of the UPA came from a variety of sources. Depending on the season, the year, and the region in which the unit was operating, as well as personal preferences, the insurgents wore a variety of clothing. The general rule among the insurgents was that they were not to wear any captured uniform in its entirety, so that there would be no possibility of confusion. Captured iconic uniform items, such as jackets and overcoats were traded, dyed in different shades and otherwise modified. Non-Ukrainian insignia would be removed and periodically the Ukrainian trident would be attached. Local craftsmen would also sew military uniforms for the UPA, at times using captured military cloth. As a result, the UPA would be clothed in a variety of items and patterns.

## Headgear

UPA insurgents wore a variety of headgear. As a light infantry force, helmets were generally considered to be a liability and rarely do they show up in the photodocumentary evidence[30]. One photograph that shows two UPA members wearing helmets also shows then brandishing bayonets at the end of their weapons, which may depict either new recruits who have not yet adopted to the ways of irregular warfare, or members with war trophies.

Figure 102. UPA insurgents wearing the UPA Mazepynka hats, courtesy of Greywolves Co. Reenactment Group.

Ukrainian popular accounts of the UPA almost exclusively depicts the fighters wearing the "Mazepa" cap (also known as a "Mazepynka"). Named for the early 18th century Ukrainian Cossack hetman Ivan Mazepa, the cap was designed in April 1914 by Dmytro Katamia and sought to update the V cut hats worn by Cossack gentry in the previous centu-

---

[30]Photographs, like documents, can be illuminating, but need to be approached with caution. It must be considered that the photographs taken by the insurgents were often posed and the fighters took time to prepare to have their pictures taken. Extra items may have been removed before a photograph was taken, while others, such as officers, may have posed with items that they usually would not be carrying. The photographs taken by the enemy, especially those of killed insurgents, may also have been, to a degree, composed.

ries. The "Mazepynka", first made popular among the Ukrainian troops in Austrian service who fought in WWI and the civil war that followed, the original had a six-panel top. A similar version of the soft cap, with its signature V in the front, was also worn by Ukrainian troops who served in the short-lived Ukrainian Carpathian Sich just before WWII. The UPA "Mazepynka" hats were sown using eight smaller panels.

Figure 103. The UPA Insurgents wearing Petlurivka hats, courtesy of Greywolves Co. Reenactment Group.

Another popular type of headgear worn by the UPA was the "Petlurivka". Named after the dynamic Ukrainian revolutionary Simon Petliura who sought to unify western and eastern Ukraine into a national republic, the "Petlurivka" was adopted by the Western Ukrainian National Republic in 1919. The cap was modeled after the WWI British forage cap but had an additional v-cut stripe added in the front. In the open space of the v, a small metal trident cockade was often attached.

Side caps were also commonly used, either modified to resemble a "Mazepynka", or worn as issued. Captured Soviet and German "kubanky" hats were also easily modified, with a V cut being added to the front of the hat. Headgear taken as was trophies was also worn by UPA insurgents at various times and places.

Cockades were commonly used by the UPA. These cockades came in different forms, but often they were made of different materials. Often the trident of the cockade was golden in color, which, most commonly, lay on a blue background. In some examples, the blue disk was surrounded by a brass guilted fluting. Other representations of the trident on a blue heraldic shield, similar to those made in Czechoslovakia before 1939, were also worn. Photographic evidence also shows that both simple metal tridents or strips of blue and white cloth were sewn into the space when cockades were not available. Side caps, either locally sewn or modified from foreign trophies, were also commonly used by the UPA. During winter, fur caps were worn by some UPA fighters.

Figure 104. Examples of Cockades worn by UPA Insurgents, from *Ukrains'ki Vijs'kovi Vidznaky*.

## Combat Uniform

No single combat uniform existed for the UPA. The item worn depending on where the insurgent found themselves as well as what year it was. In 1941, the OUN had made plans to create a Ukrainian army and developed a rather complicated uniform based on the uniform worn by soldiers of the Western Ukrainian National Republic (ZUNR) in the late 1910s. As the Nazis pushed further eastward, they began to limit Ukrainian activities and OUN activists began to dress themselves in what they could, including old Soviet and Polish uniforms. Captured German uniforms, from a variety of institutions, were also used. In 1943, Ukrainians who defected from serving in the Nazi Auxiliary Police often continued to wear their issued gear, though now stripped of Nazi emblems.

Recognizing the growing need to equip the ranks of the UPA, in 1943 the OUN leadership set up a variety of workshops. These workshops, usually located in rural villages far away from the Nazis, also provided uniforms. Often these workshops would re-work the captured materials and in some regions, dyed them in a greenish color. Buttons with tridents were manufactured in these small industrial centers, as were other metal insignia. These civilian workshops also produced and mended leather goods for the UPA, especially footwear.

Among the UPA fighters, civilian clothing, such a pants, were commonly worn, as these items wore out the quickest. In the style of the times, military tunics were often worn open, exposing the area around the neck. Photographs show that UPA fighters often wore colorful embroidered shirts under their jackets. Other items of civilian clothing also commonly appear in photographs of the UPA.

Figure 105. UPA insurgent Dmytro Bilinchuk "Cloud" wearing a traditional embroidered shirt beneath his field jacket, from *Iavorivs'kyj Fotoarkhiv UPA*.

As the Nazis began to be pushed to the west, the UPA was able to liberate large quantities of uniforms from both the Germans and their allies. Though any UPA leader who traded with the Nazis would be brought up on charges of treason and summarily executed for cooperating with the enemy, the UPA traded often with Hungarian and Italian troops making their way westward. Thus, even more varied materials began to make its way into the UPA supply chain.

By mid 1945, Soviet authorities had broken up the previous successful local production centers and, as a result, the UPA began to rely more and more on captured Soviet gear. As the previously acquired stocks of wartime material had become exhausted, the UPA insurgents had a much more difficult times to equip their forces with non-Soviet uniforms. Polish and Czechoslovakian uniform items were also used by the UPA, but as these were rather similar in appearance to Soviet uniforms, it was hard to differentiate between them. UPA security units, who began to play an ever more important role at this time, often wore the complete uniform of the enemy when conducting special operations. As the UPA abandoned large scale military operations and shifted into the shadows, the wearing of Soviet uniforms became an asset when operating in enemy controlled areas. In September 1947, when a group of thirty-six UPA insurgents successfully crossed into the American Zone, a *New York Times* article described the fighters as dressed in Soviet uniforms.[31]

---

[31] 12 September 1947, *New York Times*.

**Combat Gear**

From its inception, the UPA fighters were a mobile force of *de facto* light infantry. Though for certain missions they would use heavy equipment, such as artillery, mobility was a key feature for their survival. Non-essential equipment, or even items usually found among contemporary regular infantry units, was simply not carried. Photographic evidence, which granted, has its own limitations, usually shows an insurgent wearing a belt, to which may be attached a single strap running across the fighter's torso. Most belts appear to be made of leather and used a variety of buckles. Photographs show open buckles, similar in style to that found on a "Sam Brown Belt". German and Soviet belt buckles were also worn, but photographs and excavated examples often show that the previous totalitarian insignia was defaced or completely removed. Buckles have also been noted with prominent tridents – either added to an existing buckle or newly made.

Attached to the belt were a few items. Photographs often show insurgents wearing a pistol in a holster or a grenade. Insurgents who found themselves surrounded would often take their own lives rather than surrender to the enemy, so wearing such an item on their belt would give them easy access in a critical situation. Photographs also show a variety of knives being worn, which could have been used either in combat or for any emergency field use. As ammunition was always limited, the insurgents usually only carried a single additional ammunition pouch. Personal items, such as socks, a toothbrush, or eating utensils, were usually carried in a haversack, breadbag or map case which was traditionally worn on the hip. One photograph which shows a group of UPA insurgents on the march illustrates them carrying shelter-halves, either across their bodies or rolled up and worn on the hip.

Figure 106. UPA Insurgents on the march, from *Iavorivs'kuj Fotoarkhiv UPA*.

Officers usually carried the same items, with an additional a map cases in which a compass, pencils, maps and various documents were kept. Photographs of UPA leaders also often show them wearing binoculars around their neck. Many photographs depict senior leaders with a weapon, most often a submachinegun, on their person.

## Winter Gear

In the winter, the UPA used a variety of cold weather gear. Captured German and Soviet capes and shelter halves were often worn on patrol, especially in the rainy period. During the winter, the insurgents wore sweaters, long underwear, and gloves produced by local villagers. White camouflage clothing was also produced by villagers and was commonly worn over the standard combat uniform. At the start of 1944, for example, the UPA group "Bohun" order 1,600 sets of winter covers.[32] Long coats were also favored by the UPA, as were modified civilian cold weather coats with lamb wool collars.

Figure 107. UPA Insurgents in winter clothing, courtesy of Greywolves Co. Reenactment Group.

## Ranks and Awards

On 27 January 1944, General Order No.3/44 codified the UPA command system. For each particular function an individual fulfilled in the organization, the insurgent was to wear the appropriate symbol on their right sleeve. Thus, an insignia would show who was the Company Commander. Photos from this period also show the insignia being worn on the left sleeve and on the small part of the lapel. At about this same time, formal military ranks were adopted by the UPA and internal documents record promotions of fighters to higher ranks.

[32] Serhij Muzychuk and Ihor Marchuk, *Ukrains'ka Povstancha Armiia/Ukrainian Insurgent Army/*. Rivne: Odnostrij, 2006, p. 25.

Various groups of insurgents used particular marking to identify themselves. UPA fighters in UPA-West often adorned the lapels of their uniforms with "teeth". Made of red cloth set against a black cloth backing, these decorations were a recreation of the insignia worn by Ukrainians fighting in Western Ukraine at the end of WW1. Unlike the blue "teeth" worn in the late 1910s, those worn by UPA-West used the red and black colors of the OUN organization. Early in the war, UPA fighters wore armbands and one photo of a killed Battalion Commander from October 1944 shows the man wearing an arm patch mid-sleeve in the shape of a shield, in the center of which is a trident.

Figure 108. UPA Insurgent "Cloud" with his rank shown on his left sleeve, from *Ukrains'ka Povstancha Armiia*.

In January 1944, the UPA began to issue awards and decorations for its fighters. For those who demonstrated merit in combat, the UPA authorized a seven-tiered Merit Combat Cross and medals were awarded in Gold, Silver, and Bronze. At this time, a three-tiered Service Cross was also established, as was an award for being wounded in the line of duty. The awards could be earned by any UPA fighter, regardless of rank.

In 1948, the UPA introduced a medal for those who "Fought in exceptionally difficult circumstances". Manufactured in 1951, it is unlikely that any fighter in Ukraine ever received this metal. In April 1950, a well-known graphic designer, Nil Khasevych, designed a series of new combat awards, but it is doubtful that they were ever produced or issued. In 1952, the UPA introduced a second metal, commemorating the "10-year anniversary of the UPA. The final known awards issued in Ukraine were awarded in October 1952.

Figure 109. UPA Insurgent with "teeth" decorations on his lapels, courtesy of Greywolves Co. Reenactment Group.

Figure 110. UPA Awards, Top Row – Combat Merit Cross, Bottom Row – Merit Cross, Right Column – Medal for Combat under Circumstances of Extreme Difficulty, courtesy of Jurko Datsko.

# Appendix #3
*Figures for the Ukrainian Insurgent Army*

Despite the Second World War being among the most popular subjects in both military modeling and historical miniature wargaming, few manufactures have created figure or figure lines that move beyond the usual subjects – US, German, British, Soviet and, to a lesser extent, Japanese. The forces of smaller powers, such as France, Greece, or Poland, are much more rarely modeled, but the enterprising entrepreneur can, with a bit of effort, find figures to fill their needs. Though Ukrainian craftsmen have created one-off figures of the UPA in various scales, a gamer who is looking to field a group of Ukrainian insurgents will need to create their own representations. Those using smaller-scale figures can create approximate *ersatz* UPA troops by painting their forces a variety of uniform colors and avoiding fielding figures wearing helmets. For those using larger scale figures, a bit more creativity is required.

Figure 111. Author's 28mm UPA summer uniform force for *Bolt Action*.

The UPA used a variety of sources to supply their uniforms and weapons. These variations, discussed in the two previous appendixes, allow the gamer of creating a truly unique force. The wide scale availability of 28mm figures in both metal and plastic allows the gamer to create one off figures that can be adopted as UPA insurgents. Warlord's German and Soviet figures provide a good starting point. Both the Soviets and Nazis are represented with different body styles and it is easy to interchange parts, especially with the plastic figures. The ability to swap heads is harder with some of the metal figures, such as the Hungarian troops, but can be very rewarding. Wargame Atlantic's Partisan box, designated as French Resistance, has two bodies that can be adopted with little to no modification. The Bren gun provided in the box can represent the Czechoslovakian ZB machine gun, which was commonly used by the UPA. Other weapons in this set, such as the M3 Grease Gun and the Lee-Enfield needs to be replaced with German, Soviet, or Hungarian weapons.

Figure 112. Author's 28mm UPA winter uniform force for *Bolt Action*.

For those who game in 20mm, there are a number of figures one can modify to represent the UPA. At this scale, it is often hard to recognize individual weapon types, so this should not be a major concern for most gamers. The semi-soft plastic figures are easy to modify and while no one currently releases UPA specific figures,

Ceaser made a few sets of partisan figures in predominantly civilian dress that can be used. Strelets made a set of 1944 Finns and the majority of the figures are wearing soft caps similar to the "Mazepynka" cap worn by the UPA. A number of figures from Strelet's Hungarian Army in Winter Dress can also be adapted as UPA insurgents, as can Strelet's Romanian infantry. Of course, individual UPA insurgents can be modified from the plethora of Soviet and German figures on the market. Africa Corp figures in long pants and soft caps can work well as UPA fighters with little modification. As some units within the UPA wore reworked Italian uniforms, Italian troops can be made into UPA insurgents by doing simple head swaps to replace any troops wearing helmets. Austrian WWI figures is soft caps can be used without major modifications – by simply cutting down the rifle to carbine length, one can replicate the Hungarian carbines made from the older M95 Steyr rifles used in WWI.

# Appendix #4
*Terms and Acronyms*

**Armia Krawowa (AK)** The Polish Home Army was formed in 1942 and fought primarily against the Germans to reestablish the state of Poland within its prewar borders. In the Volyn' area, the AK fought against the UPA and the Ukrainian population. Fighting also took place between Soviet partisans and the AK. Although the AK was formally dissolved in January 1945, some units continued to fight against the Soviets into the early 1950s.

**Bandera, Stepan (1909-1959)** Controversial Ukrainian revolutionary and political leader of the OUN. In 1934, Bandera orchestrated the assignation of Poland's Minister of the Interior. Sentenced to life imprisonment, Bandera was freed when the Germans invaded Poland. Cooperation between Bandera and the Nazis was short lived and by 1942, Bandera found himself incarcerated at the Sachsenhausen Concentration Camp. Released by the Germans in the last months of the war, Bandera worked clandestinely to recreate the Ukrainian Supreme Liberation Council. In 1959, Bandera was assassinated by a KGB agent.

**Cursed Soldiers** A term used to describe the Polish resistance members who continued to fight against the Soviet authorities after the disbandment of the AK.

**Insurgents** A term used to denote military units or formations that wage war against an imperial power by engaging in asymmetric warfare. Recruited from the local population, insurgents are intimately interconnected with the indigenous population and rely on them for food, supplies, information, as well as moral support.

**Konovalets, Evhen (1891-1938)** Military commander and revolutionary leader. During WWI, Konovalets served as a Lieutenant in the Austrian Army and was captured by the Russians. Following the collapse of the Tsarist government, Konovalets organized Ukrainian POWs into a battalion of troops, which later became critical in establishing the Ukrainian National Republic. Demobilized in 1919, Konovalets was imprisoned by the Poles in Lutsk as a prisoner of war. In 1921, Konovalets founded the Ukrainian Military Organization (UVO) and in 1929, the Organization of Ukrainian Nationalists (OUN). In 1938, Konovalets was assassinated by a Soviet agent in Rotterdam.

**Melnyk, Andrij (1890-1964)** Ukrainian miliary leader and revolutionary. Melnyk served as an officer in WWI and in 1916 was taken prisoner by the Russians. During the revolution, Melnyk served as the Chief of Staff of the Army of the

Ukrainian National Republic. In 1929, he helped form the OUN and was briefly incarcerated by the Poles in the late 1920s. Considered a moderate, Melnyk was chosen to lead the OUN after Konovalet's assassination. Melnyk condemned the Nazis but cooperated with them for a short period in hopes of establishing a Ukrainian state. Arrested and imprisoned at the Sachsenhausen Concentration Camp, Melnyk survived the war and continued to work with the Ukrainian National Council.

**MVD** Russian: *Ministerstvo vnutrennikh del*/ Ministry of Internal Affairs. Formed in 1946 as the successor to the NKVD.

**NKVD** Russian: *Narodnyj komissariat vnutrennikh del*/ People's Commissariat for Internal Affairs. Renamed as the MVD in 1946.

**OUN** Organization of Ukrainian Nationalists. Formed in Vienna in 1929, the program of the organization focused on creating an independent Ukraine. Like other national liberation movements, the OUN used terrorism to achieve its goals. In 1940 the organization split into two factions, the OUN-B and the OUN-M.

**OUN-B** Organization of Ukrainian Nationalist, Bandera Faction.

**OUN-M** Organization of Ukrainian Nationalist, Melnyk Faction.

**OUN-SD** Self-Defense Units of the Organization of Ukrainian Nationalist. These were prominent before the formal creation of the UPA in 1942.

**Partisans** Irregular units that operate in the rear of an enemy during wartime. They may or may not cooperate with the local population and are provided with food and supplies from their own government.

**SB** Ukrainian: *Sluzhba Bespeky*/Security Services. These individuals were responsible for internal security with the UPA and were used to counter Soviet infiltrations of the organization.

**Ukrainian National Council (1947-1992)** Established in 1947, the Council was the parliamentary body of the Ukrainian National Republic Government in exile.

**UPA** Ukrainian Insurgent Army. Formed from OUN-SD units, the UPA was a military formation that waged asymmetric warfare on those who opposed the program of the OUN to create an independent Ukrainian state. Though many UPA units were demobilized in June 1946, the insurgents continued to wage war as part of the armed underground. The UPA command structure functioned until 1949, after which its members focused on propaganda, acts of sabotage and assassination.

**Volyn' Tragedy** Ethnic cleansing conducted in Volyn'. Volyn' was a battleground between the Polish and Ukrainian populations for much the early 20$^{th}$ century. Following the bloodbaths of 1941, specifically the wide-spread execution of prisoners by the retreating Soviet authorities and the genocide conducted against the region's Jewish population by the Nazis and their collaborators, in the years that followed the mentally brutalized and desensitizes survivors took out their frustrations on their former neighbors. Ukrainians killed Poles and Poles killed Ukrainians. While the wide-spread participation of civilians in these killings categorically rejects mono-causal explanations, at the same time one cannot dismiss the participation of organizations like the OUN-B or its members in carrying out these atrocities. Many victims were tortured, raped, and/or dismembered before being killed. Women and children were the most common victims.

Figure 113. Collective burial mound of UPA Insurgents, summer 1946, from *UPA Warfare in Ukraine*.

# Appendix #5
*List of Figures and Illustrations*

Front Cover
Small group of 28mm UPA figures, author's collection.

Figure 1. 28mm UPA figures, converted from Warlord's range of Bolt Action models, author's collection.

Figure 2. Stephan Bandera

Figure 3. Evhen Konovalets

Figure 4. Nazi-Soviet Friendship, Nazi and Soviet troops stage a joint victory parade in the city of Brest, 1939.

Figure 5. *Nachtigall* soldiers in the city of L'viv, 30 June 1941.

Figure 6. Nazis executions, 2 December 1943, Drohobych, from *UPA: They Fought Hitler and Stalin.*

Figure 7. Taras "Bulba" Borovets, 1941.

Figure 8. UPA Insurgents, 31 October 1944, from *Ukrains'ka Povstancha Armiia.*

Figure 9. Map of UPA zones, from *UPA Warfare in Ukraine.*

Figure 10. Roman Shukhevych, November 1943.

Figure 11. UPA organization, from *UPA Warfare in Ukraine.*

Figure 12. Vasyl Kuk

Figure 13. General Viktor Lutze, Bundesarchiv. B 145 Bild-F051632-0523, courtesy of Jurko Datsko.

Figure 14. Scenario 1 Map.

Figure 15. UPA Reenactor examining a German Sd.Kfz. 251 half-track, courtesy of Greywolves Co. Reenactment Group.

Figure 16. Bandai 1/48 German motorcycle and sidecar, author's collection.

Figure 17. 28mm UPA Figures, author's collection.

Figure 18. UPA insurgents from the group "Bohun" returning from a military action, 1943, from *Armia Bezsmertnykh*.

Figure 19. German troop train. Note *Flakvierling* quad AA.

Figure 20. Scenario 2 Map.

Figure 21. Group of UPA Insurgents, from *Ukrains'ka Povstans'ka Armia*.

Figure 22. 28mm German miniatures, courtesy of Matthew Doyle.

Figure 23. 28mm UPA Assault Squad, primarily armed with SVTs, author's collection.

Figure 24. UPA Reenactors, courtesy of Greywolves Co. Reenactment Group.

Figure 25. UPA Insurgents, from *Iavorivs'kyj Fotoarkhiv UPA*.

Figure 26. Area of the Slave Labor Work Camp, 2018, author's photograph.

Figure 27. UPA Insurgents, from *Iavorivs'kyj Fotoarkhiv UPA*.

Figure 28. Scenario 3 Map.

Figure 29. 28mm German camp personalities, courtesy of Warlord Games.

Figure 30. 28mm UPA figures, author's collection.

Figure 31. 28mm Hungarian troops, courtesy of Warlord Games.

Figure 32. Wooden shoe worn by a camp prisoner, 2018, author's photograph.

Figure 33. Pre-war photograph of Kamin Koshyrsky showing the town's central square, from volyntimes.com.ua.

Figure 34. Scenario 4 Map.

Figure 35. UPA insurgents in various scales, courtesy of Mykyta Karpukhin, https:www.facebook.com/karp.mykytin?mibextid+LQQJ4d.

Figure 36. 28mm German Aircrew, painted by Michael Adams, courtesy of Jon Russell.

Figure 37. Unknown UPA Insurgent, from *Ukrains'ka Povstancha Armiia*.

Figure 38. UPA reenactors, courtesy of Greywolves Co. Reenactment Group.

Figure 39. UPA NCO school taking oath, from *Ukrains'ka Povstancha Armiia*.

Figure 40. UPA reenactors, courtesy of Greywolves Co. Reenactment Group.

Figure 41. Scenario 5 map.

Figure 42. 20mm German troops, courtesy of Casey Pittman.

Figure 43. 28mm UPA sniper team, author's collection.

Figure 44. Members of the UPA "Oleni" Officer Candidate School during their Oath of Allegiance ceremony, 21 September 1944, from *UPA: They Fought Hitler and Stalin*.

Figure 45. 28mm German troops, courtesy of Matthew Doyle.

Figure 46. 1/144 German Henschel Hs 126 Reconnaissance aircraft, author's collection.

Figure 47. UPA memorial at Hurby, 2011, author's photograph.

Figure 48. Scenario 6 Map.

Figure 49. 28mm Soviet SU-76 Self-propelled gun, courtesy of Warlord Games.

Figure 50. 28mm UPA Insurgents with heavy weapons, author's collection.

Figure 51. UPA reenactors, courtesy of Greywolves Co. Reenactment Group.

Figure 52. Acting UPA commander Myroslav Symchych at the battlefield, 2 July 2012, author's photograph.

Figure 53. Western ridgeline overlooking the ambush site, 2012, author's photograph.

Figure 54. Scenario 7 Map.

Figure 55. 28mm Soviet troops, courtesy of Warlord Games.

Figure 56. 28mm UPA troops in winter camo, author's collection

Figure 57. UPA Insurgents, Birch Company, 1944, photo courtesy of Jurko Datsko.

Figure 58. 28mm Soviet partisans, courtesy of Warlord Games.

Figure 59. UPA Insurgent Mounted Reconnaissance Unit, UPA-North, Winter 1943-4, from *UPA: They Fought Hitler and Stalin*.

Figure 60. Scenario 8 Map.

Figure 61. 28mm UPA insurgents, author's collection.

Figure 62. UPA reenactors, Slavs'k, Ukraine, 2018.

Figure 63. UPA insurgents from the "Udarnyk-2" Company and US troops in West Germany, August 1947, from *Armia Bezsmertnykh*.

Figure 64. Living History Timeline display of Weapons and Uniforms used by Ukrainian Forces, Kholodnyj Iar, Ukraine, 2017, courtesy of Jurko Datsko.

Figure 65. 28mm NKVD/MVD figures, author's collection.

Figure 66. Scenario 9 Map.

Figure 67. 28mm UPA Insurgents, author's collection.

Figure 68. Personal items and field gear of UPA insurgent "Zalizniak", who in 1948 crossed into Saltsburg, Austria, from *Armia Bezsmertnykh*.

Figure 69. Plan of UPA Bunker, from *UPA Warfare in Ukraine*.

Figure 70. Living history event, Veretskyj Pass, Carpathian Mountains, 2018, courtesy of Jurko Datsko.

Figure 71. Scenario 10 Map.

Figure 72. Cover of Khrin's Memoir's, *Zymoiu v Bunkri*, author's collection.

Figure 73. 28mm UPA insurgents, author's collection.

Figure 74. Melodia the Beskid sub-district medical officer (middle) with his nurse, security and wounded, Fall 1946, from *UPA Warfare in Ukraine*.

Figure 75. 28mm Soviet NKVD/MVD figures, painted by Kostiantyn Pinaev, courtesy of Trizub Miniatures.

Figure 76. Kim Philby, 1955.

Figure 77. Airdrop.

Figure 78. Scenario 11 Map.

Figure 79. 28mm UPA couriers, author's collection.

Figure 80. 28mm Soviet NKVD Squad, courtesy of Warlord Games.

Figure 81. Kingir Camp, courtesy of Jurko Datsko.

Figure 82. Kingir Prisoners, courtesy of Jurko Datsko.

Figure 83. 54mm Soviet T-34 and troops, courtesy of Tod Kershner.

Figure 84. 28mm rioting prisoners, author's collection.

Figure 85. Scenario 12 Map.

Figure 86. 28mm Soviet NKVD/MVD figures, painted by Kostiantyn Pinaev, courtesy of Trizub Miniatures.

Figure 87. 28mm armed former prisoners, author's collection.

Figure 88. OUN cache of arms captured by the Polish Police, 1933, from *Strilets'ka*

*Zbroia Ukrains'kukh Povstantsiv.*

Figure 89. Austrian produced M95 carbine, author's collection.

Figure 90. UPA insurgents armed with SVTs, from *Iavorivs'kyj Fotoarkhiv UPA*.

Figure 91. Myroslav Symchych, on the left, with his "improved" PPS-43, courtesy of Jurko Datsko.

Figure 92. Spent Hungarian cartridges marking an insurgent's firing position were recovered during the author's Battlefield Survey of Kosmach, 2012.

Figure 93. UPA officers and men from the Black Forest group, May 1947. Note sniper rifle with PU scope on the lower righthand side of the photograph, from *Strilets'ka Zbroia Ukrains'kukh Povstantsiv*

Figure 94. UPA Insurgent with *Sturmgewehr 44*, from *Strilets'ka Zbroia Ukrains'kukh Povstantsiu*

Figure 95. Insurgents with Solothurn MG 30s, from *Ukrains'ka Povstans'ka Armia.*

Figure 96. Two UPA commanders, one armed with a Soviet submachinegun and the other with a German machine-pistol, from *UPA Warfare in Ukraine*.

Figure 97. UPA insurgents with hand grenades, from *Armia Bezsmertnykh*.

Figure 98. OUN leaders, UPA insurgents, and new recruits, 31 August 1944, from *Army of Immortals*. Note German (PzB-38 or 39?) anti-tank rifle in the front.

Figure 99. Insurgents of UPA-North with limbered artillery on the move, winter 1943-4, from *UPA: They Fought Hitler and Stalin*.

Figure 100. Captured ME 109 in Ukrainian markings, courtesy of Jurko Datsko.

Figure 101. Weapons and radios seized by Soviet authorities from a bunker, 1951, from *Strilets'ka Zbroia Ukrains'kukh Povstantsiv*. Note US M2 paratrooper carbine in the middle.

Figure 102. UPA insurgents wearing Mazepynka hats, courtesy of Greywolves Co. Reenactment Group.

Figure 103. UPA Insurgents wearing Petlurivka hats, courtesy of Greywolves Co. Reenactment Group.

Figure 104. Examples of cockades worn by UPA Insurgents, from *Ukrains'ki Vijs'kovi Vidznaky*.

Figure 105. UPA insurgent Dmytro Bilinchuk "Cloud" wearing a traditional embroidered shirt beneath his field jacket, from *Iavorivs'kyj Fotoarkhiv UPA*.

Figure 106. UPA Insurgents on the march, from *Iavorivs'kyj Fotoarkhiv UPA*.

Figure 107. UPA Insurgents in white winter clothing, courtesy of Greywolves Co. Reenactment Group.

Figure 108. UPA Insurgent "Cloud" with his rank shown on his left sleeve, from *Ukrains'ka Povstancha Armiia*.

Figure 109. UPA Insurgent with "teeth" decorations on his lapels, courtesy of Greywolves Co. Reenactment Group.

Figure 110. UPA Awards, Top Row – Combat Merit Cross, Bottom Row – Merit Cross, Right Column – Medal for Combat under Circumstances of Extreme Difficulty, courtesy of Jurko Datsko.

Figure 111. Author's 28mm UPA summer uniform force for Bolt Action.

Figure 112. Author's 28mm UPA winter uniform force for Bolt Action.

Figure 113. Collective burial mound of UPA Insurgents, summer 1946, from *UPA Warfare in Ukraine*.

Figure 114. UPA Insurgents crossing a mountain pass, from *Iavorivs'kyj Fotoarkhiv UPA*.

Back Cover
28mm UPA mortar team, author's collection

# Further Reading

Abbott, Peter and Eugene Pinak. *Ukrainian Armies 1914-1955*. Oxford: Osprey Publishing. 2004.

Bol'novs'kyj, Andrij. *Ukrains'ki vijs'kovi formuvannia v zbrojnykh sylakh Nimechchyny (1939-1945)/Ukrainian Military Formations in the German Armed Forces (1939-1945)*. L'viv: Ivan Franko L'viv National University Press. 2003.

Borovych, Ia. V. Ukraine and Poland, *Ideia i Chyn* 2, no. 4 (1943). In *Political Thought of the Ukrainian Underground, 1943-1951*. Edited by Peter J. Potichnyj and Yevhen Shtendera. Toronto: Canadian Institute of Ukrainian Studies Press. 1986.

*Iavorivs'kyj Fotoarkhiv UPA/ Iavorivs'kyj UPA Photo archive/*. L'viv: Spolom. 2005.

Ilnytzkyj, Roman, ed. „Einsatzkommando Order against the Bandera Movement, dated 25 November 1941". In *Deutschland und die Ukraine, 1934-45*, Vol. 2. Munich: Osteuropa-Institut. 1956.

Khmel', S. F. *Ukrains'ka Partyzanka/Ukrainian Insurgents/*. London: Ukrainian Publishers, Ltd. 1959.

Khrin, Stepan *Zymoiu v Bunkri /Winter in a Bunker*. Augsburg: Do Zbroi. 1950.

Konopadskyj, Oleksa *Spomyny Chotovoho Ostroverkha/Memoirs of Platoon Leader Ostroverkh*. Munich: Do Zbroi. 1953.

Kosyk, Wolodymyr. *The Third Reich and Ukraine. Studies in Modern European History*, Vol. 8. New York: Peter Lang Publishing, Inc. 1993.

Levytskyj, Myron, ed. *Istoria Ukrains'koho Vijs'ka/History of the Ukrainian Armed Forces/*. Winnipeg: Ivan Tyktor. 1953.

Mandzy, Adrian. Ambush in the Mountains: A Multi-Disciplinary Study of a Successful Field Operation by a Company of the Ukrainian Partisan Army against a Soviet NKVD Battalion in January 1945, *Partisans, Guerillas, and Irregulars: The Archaeology of Asymmetric Warfare*. Tuscaloosa: University of Alabama Press, 2019, pp. 180-199.

Manzurenko, Vitalij and Vasyl' Humeniuk. *Rejd UPA v Rumuniiu 1949 r. /UPA Raid into Romania, 1949*. L'viv: Vydavnytstvo Staroho Leva. 2007.

Marples, David. *Heroes and Villains: Creating National History on Contemporary Ukraine*. New York: Central European University Press. 2007.

Martowych, Oleh. *Ukrainian Liberation Movement in Modern Times*. Munich: Buchdruckerei Universal. [1951].

Mirchuk, Petro. *Ukrains'ka Povstans'ka Armiia 1942-1952 /Ukrainian Insurgent Army 1942-1952*. Munich: Cicero. 1953.

Muzychuk, Serhij and Ihor Marchuk. *Ukrains'ka Povstancha Armiia/Ukrainian Insurgent Army*. Rivne: Odnostrij. 2006.

Onyshchuk, Iaroslav. "Doslidzhennia pokhovannia bijtsiv Karpats'koi Ukrainy na Verets'komy perevali v 2011 rotsi/Research of the burial places of Carpathian Ukraine's soldiers on Veretsky mountain pass in 2011". *L'viv: Citadel*. No 9. 2013. pp. 46-50.

Patryliak, Ivan. *Peremoha abo Smert'/Victory or Death*. L'viv: Chasopys. 2015.

Piotrowski, Tadeusz. *Poland's Holocaust*. New York: McFarland & Co. 1998.

Posivnych, Mykola. *Voenno-polytychna diial'nist' OUN v 1929-1939/Military-political actions of the OUN in 1929-1939*. L'viv: National Academy of Sciences of Ukraine. 2010.

Semotiuk, Jaroslaw. *Ukrains'ki Vijs'kovi Vidznaky/ Ukrainian Military Medals*. Shevchenko Scientific Society in Canada. Vol. XXXIV. Toronto: Harmony Printing Limited. 1991.

Shchehliuk, Myron. *Strilets'ka Zbroia Ukrains'kukh Povstantsiv/Firearms of the Ukrainian Insurgents*. L'viv: Spolom. 2014.

Snider, Timothy. *Bloodlands: Europe between Hitler and Stalin*. New York: Basic Book. 2010.

Skorups'kyj, Maksym. *U Hastupakh I Vidstupakh/In Advance and in Retreat/*. Chicago: Ukrainian-American Publishing and Printing Co., Inc. 1961.

Sodol, Petro R. *UPA: They Fought Hitler and Stalin*. New York: Committee for the World Convention and Reunion of Soldiers in the Ukrainian Insurgent Army. 1987.

Subtelny, Orest. *Ukraine: A History*. Toronto: University of Toronto Press. 1991.

Thomas, Nigel and Peter Abbott. *Partisan Warfare 1941-45*. London: Osprey Publishing Ltd. 1983.

Tys-Krokhmaliuk, Yuriy. *UPA Warfare in Ukraine: Strategical, Tactical and Organizational Problems of Ukrainian Resistance in World War II*. New York: Society of Veterans of Ukrainian Insurgence Army. 1972.

Unknown. *The Ukrainian Insurgent Army In Fight For Freedom*. New York: Dnipro. 1953.

V'iatrovych, Volodymyr, Ruslan Zabilyj, Ihor Derev'ianyj and Petro Sodol. *Ukrains'ka Povstans'ka Armia: Istoria neskorenykh/ Ukrainian Insurgent Army: History of the Undefeated*. L'viv: Tsentr doslidzhennia vyzvol'noho rykhy. 2011.

V'iatrovych, Volodymyr and Volodymyr Moroz, ed. *Armia Bezsmertnykh: Povstans'ki Svitlyny/ Army of Immortals: Insurgent Photographs*. L'viv: Vydavnytstvo Mc. 2002.

Viatrovych, Volodymyr and Lubomyr Luciuk, ed. *Enemy Archives: Soviet Counterinsurgency Operations and the Ukrainian Nationalist Movement*. Canada: McGill-Queen's University Press. 2023.

Wallace, Samuel A. and Yaroslaw Chyz. Western Ukraine under Polish Yoke: Polonization, Colonization, and "Pacification". *The Ukrainian Review*. New York. 1931.

Western Ukraine Declares Its Independence. *Trident*. Vol. V, No. 6. July-August 1941.

Figure 114. UPA Insurgents crossing a mountain pass, from *Iavorivs'kyj Fotoarkhiv UPA*.

## About the Author

Adrian Mandzy is a Professor of History at Morehead State University and has been conducting research in Ukraine since 1989. He has taken part in archaeological excavations along the Dnister River as well as attended a graduate program in Kyiv. In 1991, Professor Mandzy organized and directed for over a decade the Kamianets-Podilsky Foundation, a non-profit organization that brought western scholars to the city of Kamianets-Podilsky.

For the last twenty years Adrian has focused on battlefield archaeology and he continues to conduct battlefield research in Ukraine, Poland, Germany and the United States. Most recently, he has been involved in archaeological digs at the Leipzig battlefield. Through the Kentucky Institute of International Studies, he directs the Slavic Europe Program, which takes college and university students to Poland and Ukraine.

Adrian has been actively involved in wargaming for over forty years. An award-winning painter, he wrote, Bad Roads and Poor Ratios: Fifty-Nine Wargame Scenarios for the North American War of 1812 for Winged Hussar Publishing. In addition, he runs Wargamer US at https://www.wargamerus.com/.

Look for more books from Winged Hussar Publishing, LLC – E-books, paperbacks and Limited-Edition hardcovers. The best in history, science fiction and fantasy at:
https://www. wingedhussarpublishing.com
https://www.whpsupplyroom.com
or follow us on Facebook at:
Winged Hussar Publishing LLC
Or on twitter at:
WingHusPubLLC
For information and upcoming publications

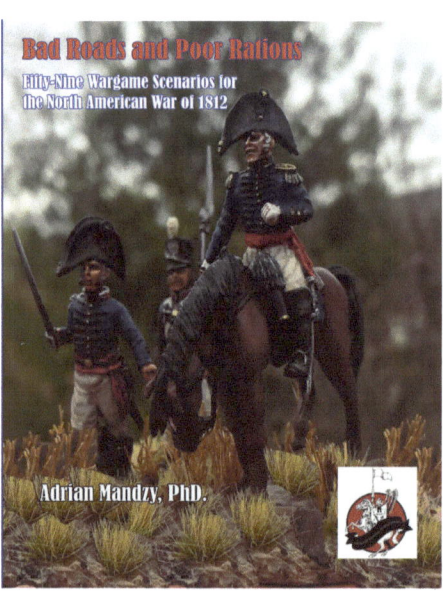

www.ingramcontent.com/pod-product-compliance
Lightning Source LLC
Chambersburg PA
CBHW040905020526
44114CB00037B/59